D0205000

Speakable and unspeakable
in quantum mechanics

WORCESTER POLYTECHNIC INST LIBRARY

To my Mother and Father

Collected papers on quantum philosophy

Speakable and unspeakable in quantum mechanics

J. S. BELL

CERN

CAMBRIDGE
UNIVERSITY PRESS

Published by the Press Syndicate of the University of Cambridge
The Pitt Building, Trumpington Street, Cambridge CB2 1RP
40 West 20th Street, New York, NY 10011–4211, USA
10 Stamford Road, Oakleigh, Melbourne 3166, Australia

© Cambridge University Press 1993

First published 1987
First paperback edition 1988
Reprinted 1989 (twice), 1991, 1993

Printed in Great Britain at the University Press, Cambridge

British Library cataloguing in publication data
Bell, J. S.
Speakable and unspeakable in quantum
mechanics: collected papers in quantum
mechanics.

1. Quantum theory
I. Title
530.1'2   QC174.12

Library of Congress cataloguing in publication data
Bell, J. S.
Speakable and unspeakable in quantum mechanics.
Includes index.
1. Quantum theory – Collected works.   I. Title.
QC173.97.B45 1987   530.1'2   86-32728

ISBN 0 521 33495 0 hardback
ISBN 0 521 36869 3 paperback

TM

Contents

QC
173.97
B45
1987

J. S. Bell: Papers on quantum philosophy

On the problem of hidden variables in quantum mechanics. *Reviews of Modern Physics* **38** (1966) 447–52.

On the Einstein–Podolsky–Rosen paradox. *Physics* **1** (1964) 195–200.

The moral aspect of quantum mechanics. (with M. Nauenberg) In *Preludes in Theoretical Physics*, edited by A. De Shalit, H. Feshbach, and L. Van Hove. North Holland, Amsterdam, (1966) pp 279–86.

Introduction to the hidden-variable question. *Foundations of Quantum Mechanics.* Proceedings of the International School of Physics 'Enrico Fermi', course IL, New York, Academic (1971) pp 171–81.

On the hypothesis that the Schrödinger equation is exact. TH-1424-CERN October 27, 1971. Contribution to the International Colloquium on Issues in Contemporary Physics and Philosopy of Science, and their Relevance for our Society, Penn State University, September 1971. Reproduced in *Epistemological Letters*, July 1978, pp 1–28, and here in revised form as 15. Omitted.

Subject and Object. In *The Physicist's Conception of Nature* Dordrecht-Holland, D. Reidel (1973) pp 687–90.

On wave packet reduction in the Coleman–Hepp model. *Helvetica Physica Acta* **48** (1975) 93–8.

The theory of local beables. TH-2053-CERN, 1975 July 28. Presented at the Sixth GIFT Seminar, Jaca, 2–7 June 1975, and reproduced in *Epistemological Letters*, March 1976.

Locality in quantum mechanics: reply to critics. *Epistemological Letters*, Nov. 1975, pp 2–6.

How to teach special relativity. *Progress in Scientific Culture*, Vol 1, No 2, summer 1976.

Einstein–Podolsky–Rosen experiments. *Proceedings of the Symposium on Frontier Problems in High Energy Physics*, Pisa, June 1976, pp 33–45.

The measurement theory of Everett and de Broglie's pilot wave. In *Quantum Mechanics, Determinism, Causality, and Particles*, edited by M. Flato *et al.* Dordrecht-Holland, D. Reidel, (1976) pp 11–17.

Free variables and local causality. *Epistemological Letters*, Feb. 1977.

Atomic-cascade photons and quantum-mechanical nonlocality. *Comments on Atomic and Molecular Physics* 9 (1980) pp 121–6. Invited talk at the Conference of the European Group for Atomic spectroscopy, Orsay-Paris, 10–13 July, 1979.

de Broglie-Bohm, delayed-choice double-slit experiment, and density matrix. *International Journal of Quantum Chemistry*: Quantum Chemistry Symposium 14 (1980) 155–9.

Quantum mechanics for cosmologists. In *Quantum Gravity* 2, editors C. Isham, R. Penrose, and D. Sciama. Oxford, Clarendon Press (1981) pp 611–37. Revised version of 'On the hypothesis that the Schrödinger equation is exact' (see above).

Bertlmann's socks and the nature of reality. *Journal de Physique*, Colloque C2, suppl. au numero 3, Tome 42 (1981) pp C2 41–61.

On the impossible pilot wave. *Foundations of Physics* 12 (1982) pp 989–99.

Speakable and unspeakable in quantum mechanics. Introductory remarks at Naples–Amalfi meeting, May 7, 1984.

Quantum field theory without observers. Talk at Naples–Amalfi meeting, May 11, 1984. (Preliminary version of 'Beables for quantum field theory'.) Omitted.

Beables for quantum field theory. 1984 Aug 2, CERN-TH. 4035/84.

Six possible worlds of quantum mechanics. *Proceedings of the Nobel Symposium 65: Possible Worlds in Arts and Sciences*. Stockholm, August 11–15, 1986.

EPR correlations and EPW distributions. In *New Techniques and Ideas in Quantum Measurement Theory* (1986). New York Academy of Sciences.

Are there quantum jumps? In *Schrödinger. Centenary of a polymath* (1987). Cambridge University Press.

Preface

Simon Capelin, of Cambridge University Press, suggested that I send him my papers on quantum philosophy and let him make them into a book. I have done so. The papers, from the years 1964–1986, are presented here in the order, as far as I now can tell, in which they were written. But of course that is not the order, if any, in which they should be read.

Papers 18 and 20, 'Speakable and unspeakable in quantum mechanics' and 'Six possible worlds of quantum mechanics', are nontechnical introductions to the subject. They are meant to be intelligible to nonphysicists. So also is most of paper 16, 'Bertlmann's socks and the nature of reality', which is concerned with the problem of apparent action at a distance.

For those who know something of quantum formalism, paper 3, 'The moral aspect of quantum mechanics', introduces the infamous 'measurement problem'. I thank Michael Nauenberg, who was co-author of that paper, for permission to include it here. At about the same level, paper 17, 'On the impossible pilot wave', begins the discussion of 'hidden variables', and of related 'impossibility' proofs.

More elaborate discussions of the 'measurement problem' are given in paper 6, 'On wavepacket reduction in the Coleman–Hepp model', and in 15, 'Quantum mechanics for cosmologists'. These show my conviction that, despite numerous solutions of the problem 'for all practical purposes', a problem of principle remains. It is that of locating precisely the boundary between what must be described by wavy quantum states on the one hand, and in Bohr's 'classical terms' on the other. The elimination of this shifty boundary has for me always been the main attraction of the 'pilot-wave' picture.

Of course, despite the unspeakable 'impossibility proofs', the pilot-wave picture of de Broglie and Bohm exists. Moreover, in my opinion, all students should be introduced to it, for it encourages flexibility and precision of thought. In particular, it illustrates very explicitly Bohr's insight that the result of a 'measurement' does not in general reveal some preexisting property of the 'system', but is a product of both 'system' and 'apparatus'. It seems to me that full appreciation of this would have aborted most of the 'impossibility proofs', and most of 'quantum logic'. Papers 1 and 4, as well as 17, dispose of 'impossibility proofs'. More constructive expositions of various aspects of the pilot-wave picture are contained in papers 1, 4, 11, 14, 15, 17, and 19. Most of this is for nonrelativistic quantum mechanics, but the last paper, 19, 'Beables for quantum field theory', discusses relativistic extensions. While the usual predictions are obtained for experimental tests of special relativity, it is lamented that a preferred frame of reference is involved behind the phenomena. In this connection one paper, 9, 'How to teach special relativity', has been included although it has no particular reference to quantum mechanics. I think that it may be helpful as regards the preferred frame, at the fundamental level, in 19. Many students never realize, it seems to me, that this primitive attitude, admitting a special system of reference which is experimentally inaccessible, is consistent... if unsophisticated.

Any study of the pilot-wave theory, when more than one particle is considered, leads quickly to the question of action at a distance, or 'nonlocality', and the Einstein–Podolsky–Rosen correlations. This is considered briefly in several of the papers already mentioned, and is the main concern of most of the others. On this question I suggest that even quantum experts might begin with 16, 'Bertlmann's socks and the nature of reality', not skipping the slightly more technical material at the end. Seeing again what I have written on the locality business, I regret never having written up the version of the locality inequality theorem that I have been mostly using in talks on this subject in recent years. But the reader can easily reconstruct that. It begins by emphasizing the need for the concept 'local beable', along the lines of the introduction to 7. (If local causality in some theory is to be examined, then one must decide which of the many mathematical entities that appear are supposed to be real, and really here rather than there). Then the simpler locality condition appended to 21 is formulated (rather than the more elaborate condition of 7). With an argument modelled on that of 7 the factorization of the probability distribution again follows. The Clauser–Holt–Horne–Shimony inequality is then obtained as at the end of 16.

My attitude to the Everett–de Witt 'many world' interpretation, a rather negative one, is set out in paper 11, 'The measurement theory of Everett and de Broglie's pilot wave', and in 15, 'Quantum mechanics for cosmologists'. There are also some remarks in paper 20.

There is much overlap between the papers. But the fond author can see something distinctive in each. I could bring myself to omit only a couple

which were used again later with slight modifications. The later versions are
included as 15 and 19.

For reproduction here, some trivial slips have been corrected, and
references to preprints have been replaced by references to publications
where possible.

In the individual papers I have thanked many colleagues for their help.
But I here renew very especially my warm thanks to Mary Bell. When I look
through these papers again I see her everywhere.

J. S. Bell, Geneva, March, 1987.

Acknowledgements

1   On the problem of hidden variables in quantum theory. *Rev. Mod. Phys.* **38** (1966) 447–52. Reprinted by permission of The American Physical Society.

2   On the Einstein–Podolsky–Rosen paradox. *Physics* **1** (1964) 195–200. Reprinted by permission of The American Physical Society.

3   The moral aspect of quantum mechanics. (with M. Nauenberg) In *Preludes in Theoretical Physics*, edited by A. De Shalit, H. Feshbach, and L. Van Hove, North Holland, Amsterdam (1966) 279–86. Reprinted by permission of North-Holland Physics Publishing, Amsterdam.

4   Introduction to the hidden-variable question. *Proceedings of the International School of Physics 'Enrico Fermi', course IL: Foundations of Quantum Mechanics.* New York, Academic (1971) 171–81. Reprinted by permission of Societa Italiana di Fisica.

5   Subject and Object. In *The Physicist's Conception of Nature*, edited by J. Mehra. D. Reidel, Dordrecht, Holland, (1973) 687–90. Copyright © 1973 by D. Reidel Publishing Company, Dordrecht, Holland.

6   On wave packet reduction in the Coleman–Hepp model. *Helvetica Physica Acta*, **48** (1975) 93–8. Reprinted by permission of Birkhauser Verlag, Basel.

7   The theory of local beables. TH-2053-CERN, 1975 July 28. Presented at the sixth GIFT seminar, Jaca, 2–7 June 1975, and reproduced in Epistemological Letters March 1976. Reprinted by permission of the Association Ferdinand Gonseth. This article also appeared in Dialectica 39 (1985) 86.

8   Locality in quantum mechanics: reply to critics. Epistemological Letters, Nov. 1975, 2–6. Reprinted by permission of the Association Ferdinand Gonseth.

9   How to teach special relativity. *Progress in Scientific Culture*, Vol 1, No 2, summer 1976. Reprinted by permission of the Ettore Majorana Centre.

10 Einstein–Podolsky–Rosen experiments. *Proceedings of the symposium on Frontier Problems in High Energy Physics.* Pisa, June 1976, 33–45. Reprinted by permission of the Annali Della Schola Normale Superiore di Pisa.

11 The measurement theory of Everett and de Broglie's pilot wave. In *Quantum Mechanics, Determinism, Causality, and Particles*, edited by M. Flato et al. D. Reidel, Dordrecht, Holland, (1976) 11–17. Copyright © 1976 by D. Reidel Publishing Company, Dordrecht, Holland.

12   Free variables and local causality. Epistemological Letters, February 1977. Reprinted by permission of Association Ferdinand Gonseth. This article also appeared in Dialectica 39 (1985) 103.

13   Atomic-cascade photons and quantum-mechanical nonlocality. *Comments on atomic and Molecular Physics* **9** (1980) 121–26. Invited talk at the Conference of the European Group for Atomic spectroscopy, Orsay-Paris, 10–13 July, 1979. Reprinted by permission of the author and publishers. Copyright © Gordon and Breach Science Publishers, Inc.

14   de Broglie-Bohm, delayed-choice double-slit experiment, and density matrix. *Internationl Journal of Quantum Chemistry*: Quantum Chemistry Symposium 14 (1980) 155–59. Copyright © 1980 John Wiley and Sons. Reprinted by permission of John Wiley and Sons, Inc.

15   Quantum mechanics for cosmologists. In *Quantum Gravity* **2**, editors C. Isham, R. Penrose, and D. Sciama, Clarendon Press, Oxford (1981) 611–37. Reprinted by permission of Oxford University Press.

16   Bertlmann's socks and the nature of reality. *Journal de Physique*, Colloque C2, suppl. au numero 3, Tome 42 (1981) C2 41–61. Reprinted by permission of Les Editions de Physique.

17   On the impossible pilot wave. *Foundations of Physics* **12** (1982) 989–99. Reprinted by permission of Plenum Publishing Corporation.

18   Beables for quantum field theory. 1984 Aug 2, CERN-TH.4035/84. Reprinted by permission of Routledge & Kegan Paul.

19   Six possible worlds of quantum mechanics. *Proceedings of the Noble Symposium 65: Possible Worlds in Arts and Sciences*. Stockholm, August 11–15, 1986, edited by Sture Allén. Reprinted by permission of The Nobel Foundation.

20   EPR correlations and EPW distributions. *In New Techniques and Ideas in Quantum Measurement Theory* (1986). Reprinted by permission of the New York Academy of Sciences.

1

On the problem of hidden variables in quantum mechanics*

1 Introduction

To know the quantum mechanical state of a system implies, in general, only statistical restrictions on the results of measurements. It seems interesting to ask if this statistical element be thought of as arising, as in classical statistical mechanics, because the states in question are averages over better defined states for which individually the results would be quite determined. These hypothetical 'dispersion free' states would be specified not only by the quantum mechanical state vector but also by additional 'hidden variables' – 'hidden' because if states with prescribed values of these variables could actually be prepared, quantum mechanics would be observably inadequate.

Whether this question is indeed interesting has been the subject of debate.[1,2] The present paper does not contribute to that debate. It is addressed to those who do find the question interesting, and more particularly to those among them who believe that[3] 'the question concerning the existence of such hidden variables received an early and rather decisive answer in the form of von Neumann's proof on the mathematical impossibility of such variables in quantum theory.' An attempt will be made to clarify what von Neumann and his successors actually demonstrated. This will cover, as well as von Neumann's treatment, the recent version of the argument by Jauch and Piron,[3] and the stronger result consequent on the work of Gleason.[4] It will be urged that these analyses leave the real question untouched. In fact it will be seen that these demonstrations require from the hypothetical dispersion free states, not only that appropriate ensembles thereof should have all measurable properties of quantum mechanical states, but certain other properties as well. These additional demands appear reasonable when results of measurement are loosely identified with properties of isolated systems.

* Work supported by U.S. Atomic Energy Commission. *Stanford Linear Accelerator Center, Stanford University, Stanford, California.*

They are seen to be quite unreasonable when one remembers with Bohr[5] 'the impossibility of any sharp distinction between the behaviour of atomic objects and the interaction with the measuring instruments which serve to define the conditions under which the phenomena appear.'

The realization that von Neumann's proof is of limited relevance has been gaining ground since the 1952 work of Bohm.[6] However, it is far from universal. Moreover, the writer has not found in the literature any adequate analysis of what went wrong.[7] Like all authors of noncommissioned reviews, he thinks that he can restate the position with such clarity and simplicity that all previous discussions will be eclipsed.

2 Assumptions, and a simple example

The authors of the demonstrations to be reviewed were concerned to assume as little as possible about quantum mechanics. This is valuable for some purposes, but not for ours. We are interested only in the possibility of hidden variables in ordinary quantum mechanics and will use freely all the usual notions. Thereby the demonstrations will be substantially shortened.

A quantum mechanical 'system' is supposed to have 'observables' represented by Hermitian operators in a complex linear vector space. Every 'measurement' of an observable yields one of the eigenvalues of the corresponding operator. Observables with commuting operators can be measured simultaneously.[8] A quantum mechanical 'state' is represented by a vector in the linear state space. For a state vector ψ the statistical expectation value of an observable with operator O is the normalized inner product $(\psi, O\psi)/(\psi, \psi)$.

The question at issue is whether the quantum mechanical states can be regarded as ensembles of states further specified by additional variables, such that given values of these variables together with the state vector determine precisely the results of individual measurements. These hypothetical well-specified states are said to be 'dispersion free.'

In the following discussion it will be useful to keep in mind as a simple example a system with a two-dimensional state space. Consider for definiteness a spin $-\frac{1}{2}$ particle without translational motion. A quantum mechanical state is represented by a two-component state vector, or spinor, ψ. The observables are represented by 2×2 Hermitian matrices

$$\alpha + \boldsymbol{\beta} \cdot \boldsymbol{\sigma}, \tag{1}$$

where α is a real number, $\boldsymbol{\beta}$ a real vector, and $\boldsymbol{\sigma}$ has for components the Pauli matrices; α is understood to multiply the unit matrix. Measurement of such an observable yields one of the eigenvalues.

$$\alpha \pm |\boldsymbol{\beta}|, \qquad (2)$$

with relative probabilities that can be inferred from the expectation value

$$\langle \alpha + \boldsymbol{\beta}\cdot\boldsymbol{\sigma} \rangle = (\psi, [\alpha + \boldsymbol{\beta}\cdot\boldsymbol{\sigma}]\psi).$$

For this system a hidden variable scheme can be supplied as follows: The dispersion free states are specified by a real number λ, in the interval $-\frac{1}{2} \leqslant \lambda \leqslant \frac{1}{2}$, as well as the spinor ψ. To describe how λ determines which eigenvalue the measurement gives, we note that by a rotation of coordinates ψ can be brought to the form

$$\psi = \begin{pmatrix} 1 \\ 0 \end{pmatrix}.$$

Let $\beta_x, \beta_y, \beta_z$, be the components of $\boldsymbol{\beta}$ in the new coordinate system. Then measurement of $\alpha + \boldsymbol{\beta}\cdot\boldsymbol{\sigma}$ on the state specified by ψ and λ results with certainty in the eigenvalue

$$\alpha + |\boldsymbol{\beta}|\,\mathrm{sign}\,(\lambda|\boldsymbol{\beta}| + \tfrac{1}{2}|\beta_z|)\,\mathrm{sign}\,X, \qquad (3)$$

where

$$
\begin{aligned}
X &= \beta_z && \text{if } \beta_z \neq 0 \\
&= \beta_x && \text{if } \beta_z = 0, \quad \beta_x \neq 0 \\
&= \beta_y && \text{if } \beta_z = 0, \quad \text{and } \beta_x = 0
\end{aligned}
$$

and

$$
\begin{aligned}
\mathrm{sign}\,X &= +1 && \text{if } X \geqslant 0 \\
&= -1 && \text{if } X < 0.
\end{aligned}
$$

The quantum mechanical state specified by ψ is obtained by uniform averaging over λ. This gives the expectation value

$$\langle \alpha + \boldsymbol{\beta}\cdot\boldsymbol{\sigma} \rangle = \int_{-1/2}^{1/2} d\lambda\{\alpha + |\boldsymbol{\beta}|\,\mathrm{sign}\,(\lambda|\boldsymbol{\beta}| + \tfrac{1}{2}|\beta_z|)\,\mathrm{sign}\,X\} = \alpha + \beta_z$$

as required.

It should be stressed that no physical significance is attributed here to the parameter λ and that no pretence is made of giving a complete reinterpretation of quantum mechanics. The sole aim is to show that at the level considered by von Neumann such a reinterpretation is not excluded. A complete theory would require for example an account of the behaviour of the hidden variables during the measurement process itself. With or without hidden variables the analysis of the measurement process presents peculiar difficulties,[8] and we enter upon it no more than is strictly necessary for our very limited purpose.

3 von Neumann

Consider now the proof of von Neumann[9] that dispersion free states, and so hidden variables, are impossible. His essential assumption[10] is: *Any real linear combination of any two Hermitian operators represents an observable, and the same linear combination of expectation values is the expectation value of the combination.* This is true for quantum mechanical states; it is required by von Neumann of the hypothetical dispersion free states also. In the two-dimensional example of Section 2, the expectation value must then be a linear function of α and β. But for a dispersion free state (which has no statistical character) the expectation value of an observable must equal one of its eigenvalues. The eigenvalues (2) are certainly not linear in β. Therefore, dispersion free states are impossible. If the state space has more dimensions, we can always consider a two-dimensional subspace; therefore, the demonstration is quite general.

The essential assumption can be criticized as follows. At first sight the required additivity of expectation values seems very reasonable, and it is rather the non-additivity of allowed values (eigenvalues) which requires explanation. Of course the explanation is well known: A measurement of a sum of noncommuting observables cannot be made by combining trivially the results of separate observations on the two terms – it requires a quite distinct experiment. For example the measurement of σ_x for a magnetic particle might be made with a suitably oriented Stern–Gerlach magnet. The measurement of σ_y would require a different orientation, and of $(\sigma_x + \sigma_y)$ a third and different orientation. But this explanation of the nonadditivity of allowed values also established the nontriviality of the additivity of expectation values. The latter is a quite peculiar property of quantum mechanical states, not to be expected *a priori*. There is no reason to demand it individually of the hypothetical dispersion free states, whose function it is to reproduce the *measurable* peculiarities of quantum mechanics *when averaged over.*

In the trivial example of Section 2 the dispersion free states (specified λ) have additive expectation values only for commuting operators. Nevertheless, they give logically consistent and precise predictions for the results of all possible measurements, which when averaged over λ are fully equivalent to the quantum mechanical predictions. In fact, for this trivial example, the hidden variable question as posed informally by von Neumann[11] in his book is answered in the affirmative.

Thus the formal proof of von Neumann does not justify his informal conclusion[12]: 'It is therefore not, as is often assumed, a question of reinterpretation of quantum mechanics – the present system of quantum mechanics would have to be objectively false in order that another description of the elementary process than the statistical one be possible.' It was not the objective measurable predictions of quantum mechanics which ruled out hidden variables. It was the arbitrary assumption of a particular (and impossible) relation between the results of incompatible measurements either of which *might* be made on a given occasion but only one of which can in fact be made.

4 Jauch and Piron

A new version of the argument has been given by Jauch and Piron.[3] Like von Neumann they are interested in generalized forms of quantum mechanics and do not assume the usual connection of quantum mechanical expectation values with state vectors and operators. We assume the latter and shorten the argument, for we are concerned here only with possible interpretations of ordinary quantum mechanics.

Consider only observables represented by projection operators. The eigenvalues of projection operators are 0 and 1. Their expectation values are equal to the probabilities that 1 rather than 0 is the result of measurement. For any two projection operators, a and b, a third $(a \cap b)$ is defined as the projection on to the intersection of the corresponding subspaces. The essential axioms of Jauch and Piron are the following:

(A) Expectation values of *commuting* projection operators are additive.

(B) If, for some state and two projections a and b,

$$\langle a \rangle = \langle b \rangle = 1,$$

then for that state

$$\langle a \cap b \rangle = 1.$$

Jauch and Piron are led to this last axiom ($4°$ in their numbering) by an analogy with the calculus of propositions in ordinary logic. The projections are to some extent analogous to logical propositions, with the allowed value 1 corresponding to 'truth' and 0 to 'falsehood,' and the construction $(a \cap b)$ to (a 'and' b). In logic we have, of course, if a is true and b is true then (a and b) is true. The axiom has this same structure.

Now we can quickly rule out dispersion free states by considering a two-dimensional subspace. In that the projection operators are the zero, the unit operator, and those of the form

$$\tfrac{1}{2} + \tfrac{1}{2}\hat{\alpha}\cdot\sigma,$$

where $\hat{\alpha}$ is a unit vector. In a dispersion free state the expectation value of an operator must be one of its eigenvalues, 0 or 1 for projections. Since from (A)

$$\langle \tfrac{1}{2} + \tfrac{1}{2}\hat{\alpha}\cdot\boldsymbol{\sigma} \rangle + \langle \tfrac{1}{2} - \tfrac{1}{2}\hat{\alpha}\cdot\boldsymbol{\sigma} \rangle = 1,$$

we have that for a dispersion free state either

$$\langle \tfrac{1}{2} + \tfrac{1}{2}\hat{\alpha}\cdot\boldsymbol{\sigma} \rangle = 1 \quad \text{or} \quad \langle \tfrac{1}{2} - \tfrac{1}{2}\hat{\alpha}\cdot\boldsymbol{\sigma} \rangle = 1.$$

Let $\hat{\alpha}$ and $\hat{\beta}$ be any noncollinear unit vectors and

$$a = \tfrac{1}{2} \pm \tfrac{1}{2}\hat{\alpha}\cdot\boldsymbol{\sigma}, \qquad b = \tfrac{1}{2} \pm \tfrac{1}{2}\hat{\beta}\cdot\boldsymbol{\sigma},$$

with the signs chosen so that $\langle a \rangle = \langle b \rangle = 1$. Then (B) requires

$$\langle a \cap b \rangle = 1.$$

But with $\hat{\alpha}$ and $\hat{\beta}$ noncollinear, one readily sees that

$$a \cap b = 0,$$

so that

$$\langle a \cap b \rangle = 0.$$

So there can be no dispersion free states.

The objection to this is the same as before. We are not dealing in (B) with logical propositions, but with measurements involving, for example, differently oriented magnets. The axiom holds for quantum mechanical states.[13] But it is a quite peculiar property of them, in no way a necessity of thought. Only the quantum mechanical averages over the dispersion free states need reproduce this property, as in the example of Section 2.

5 Gleason

The remarkable mathematical work of Gleason[4] was not explicitly addressed to the hidden variable problem. It was directed to reducing the axiomatic basis of quantum mechanics. However, as it apparently enables von Neumann's result to be obtained without objectionable assumptions about noncommuting operators, we must clearly consider it. The relevant corollary of Gleason's work is that, if the dimensionality of the state space is greater than two, the additivity requirement for expectation values of *commuting operators* cannot be met by dispersion free states. This will now be proved, and then its significance discussed. It should be stressed that Gleason obtained more than this, by a lengthier argument, but this is all that is essential here.

It suffices to consider projection operators. Let $P(\phi)$ be the projector on to the Hilbert space vector ϕ, i.e., acting on any vector ψ

$$P(\phi)\psi = (\phi, \phi)^{-1}(\phi, \psi)\phi.$$

If a set ϕ_i are complete and orthogonal,

$$\sum_i P(\phi_i) = 1.$$

Since the $P(\phi_i)$ commute, by hypothesis then

$$\sum_i \langle P(\phi_i) \rangle = 1. \tag{4}$$

Since the expectation value of a projector is non-negative (each measurement yields one of the allowed values 0 or 1), and since any two orthogonal vectors can be regarded as members of a complete set, we have:

(A) If with some vector ϕ, $\langle P(\phi) \rangle = 1$ for a given state, then for that state $\langle P(\psi) \rangle = 0$ for any ψ orthogonal on ϕ.

If ψ_1 and ψ_2 are another orthogonal basis for the subspace spanned by some vectors ϕ_1 and ϕ_2, then from (4)

$$\langle P(\psi_1) \rangle + \langle P(\psi_2) \rangle = 1 - \sum_{i \neq 1, i \neq 2} \langle P(\phi_i) \rangle$$

or

$$\langle P(\psi_1) \rangle + \langle P(\psi_2) \rangle = \langle P(\phi_1) \rangle + \langle P(\phi_2) \rangle.$$

Since ψ_1 may be any combination of ϕ_1 and ϕ_2, we have:

(B) If for a given state

$$\langle P(\phi_1) \rangle = \langle P(\phi_2) \rangle = 0$$

for some pair of orthogonal vectors, then

$$\langle P(\alpha\phi_1 + \beta\phi_2) \rangle = 0$$

for all α and β.

(A) and (B) will now be used repeatedly to establish the following. Let ϕ and ψ be some vectors such that for a given state

$$\langle P(\psi) \rangle = 1, \tag{5}$$

$$\langle P(\phi) \rangle = 0. \tag{6}$$

Then ϕ and ψ cannot be arbitrarily close; in fact

$$|\phi - \psi| > \tfrac{1}{2}|\psi|. \tag{7}$$

To see this let us normalize ψ and write ϕ in the form

$$\phi = \psi + \varepsilon\psi',$$

where ψ' is orthogonal to ψ and normalized and ε is a real number. Let ψ'' be a normalized vector orthogonal to both ψ and ψ' (it is here that we need three dimensions at least) and so to ϕ. By (A) and (5),

$$\langle P(\psi')\rangle = 0, \qquad \langle P(\psi'')\rangle = 0.$$

Then by (B) and (6),

$$\langle P(\phi + \gamma^{-1}\varepsilon\psi'')\rangle = 0,$$

where γ is any real number, and also by (B),

$$\langle P(-\varepsilon\psi' + \gamma\varepsilon\psi'')\rangle = 0.$$

The vector arguments in the last two formulas are orthogonal; so we may add them, again using (B):

$$\langle P(\psi + \varepsilon(\gamma + \gamma^{-1})\psi'')\rangle = 0.$$

Now if ε is less than $\frac{1}{2}$, there are real γ such that

$$\varepsilon(\gamma + \gamma^{-1}) = \pm 1.$$

Therefore,

$$\langle P(\psi + \psi'')\rangle = \langle P(\psi - \psi'')\rangle = 0.$$

The vectors $\psi \pm \psi''$ are orthogonal; adding them and again using (B),

$$\langle P(\psi)\rangle = 0.$$

This contradicts the assumption (5). Therefore,

$$\varepsilon > \tfrac{1}{2},$$

as announced in (7).

Consider now the possibility of dispersion free states. For such states each projector has expectation value either 0 or 1. It is clear from (4) that both values must occur, and since there are no other values possible, there must be arbitrarily close pairs ψ, ϕ with different expectation values 0 and 1, respectively. But we saw above such pairs could *not* be arbitrarily close. Therefore, there are no dispersion free states.

That so much follows from such apparently innocent assumptions leads us to question their innocence. Are the requirements imposed, which are satisfied by quantum mechanical states, reasonable requirements on the dispersion free states? Indeed they are not. Consider the statement (B). The operator $P(\alpha\phi_1 + \beta\phi_2)$ commutes with $P(\phi_1)$ and $P(\phi_2)$ only if either α or β is zero. Thus in general measurement of $P(\alpha\phi_1 + \beta\phi_2)$ requires a quite distinct experimental arrangement. We can therefore reject (B) on the grounds already used: it relates in a nontrivial way the results of experiments which cannot be performed simultaneously; the dispersion free states need not have this property, it will suffice if the quantum mechanical averages over them do. How did it come about that (B) was a consequence of assumptions in which only commuting operators were explicitly mentioned? The danger in fact was not in the explicit but in the implicit assumptions. It was tacitly assumed that measurement of an observable must yield the same value independently of what other measurements may be made simultaneously. Thus as well as $P(\phi_3)$ say, one might measure *either* $P(\phi_2)$ *or* $P(\psi_2)$, where ϕ_2 and ψ_2 are orthogonal to ϕ_3 but not to one another. These different possibilities require different experimental arrangements; there is no *a priori* reason to believe that the results for $P(\phi_3)$ should be the same. The result of an observation may reasonably depend not only on the state of the system (including hidden variables) but also on the complete disposition of the apparatus; see again the quotation from Bohr at the end of Section 1.

To illustrate these remarks, we construct a very artificial but simple hidden variable decomposition. If we regard all observables as functions of commuting projectors, it will suffice to consider measurements of the latter. Let $P_1, P_2, \cdots$ be the set of projectors measured by a given apparatus, and for a given quantum mechanical state let their expectation values be λ_1, $\lambda_2 - \lambda_1, \lambda_3 - \lambda_2, \cdots$. As hidden variable we take a real number $0 < \lambda \leqslant 1$; we specify that measurement on a state with given λ yields the value 1 for P_n if $\lambda_{n-1} < \lambda \leqslant \lambda_n$, and zero otherwise. The quantum mechanical state is obtained by uniform averaging over λ. There is no contradiction with Gleason's corollary, because the result for a given P_n depends also on the choice of the others. Of course it would be silly to let the result be affected by a mere permutation of the other Ps, so we specify that the same order is taken (however defined) when the Ps are in fact the same set. Reflection will deepen the initial impression of artificiality here. However, the example suffices to show that the implicit assumption of the impossibility proof was essential to its conclusion. A more serious hidden variable decomposition will be taken up in Section 6.[14]

6 Locality and separability

Up till now we have been resisting arbitrary demands upon the hypothetical dispersion free states. However, as well as reproducing quantum mechanics on averaging, there *are* features which can reasonably be desired in a hidden variable scheme. The hidden variables should surely have some spacial significance and should evolve in time according to prescribed laws. These are prejudices, but it is just this possibility of interpolating some (preferably causal) space-time picture, between preparation of and measurements on states, that makes the quest for hidden variables interesting to the unsophisticated.[2] The ideas of space, time, and causality are not prominent in the kind of discussion we have been considering above. To the writer's knowledge the most successful attempt in that direction is the 1952 scheme of Bohm for elementary wave mechanics. By way of conclusion, this will be sketched briefly, and a curious feature of it stressed.

Consider for example a system of two spin $-\frac{1}{2}$ particles. The quantum mechanical state is represented by a wave function,

$$\psi_{ij}(\mathbf{r}_1, \mathbf{r}_2),$$

where i and j are spin indices which will be suppressed. This is governed by the Schrödinger equation,

$$\partial\psi/\partial t = -i(-(\partial^2/\partial\mathbf{r}_1^2) - (\partial^2/\partial\mathbf{r}_2^2) + V(\mathbf{r}_1 - \mathbf{r}_2)$$
$$+ a\boldsymbol{\sigma}_1 \cdot \mathbf{H}(\mathbf{r}_1) + b\boldsymbol{\sigma}_2 \cdot \mathbf{H}(\mathbf{r}_2))\psi, \tag{8}$$

where V is the interparticle potential. For simplicity we have taken neutral particles with magnetic moments, and an external magnetic field $\mathbf{H}$ has been allowed to represent spin analyzing magnets. The hidden variables are then two vectors $\mathbf{X}_1$ and $\mathbf{X}_2$, which give directly the results of position measurements. Other measurements are reduced ultimately to position measurements.[15] For example, measurement of a spin component means observing whether the particle emerges with an upward or downward deflection from a Stern–Gerlach magnet. The variables $\mathbf{X}_1$ and $\mathbf{X}_2$ are supposed to be distributed in configuration space with the probability density,

$$\rho(\mathbf{X}_1, \mathbf{X}_2) = \sum_{ij} |\psi_{ij}(\mathbf{X}_1, \mathbf{X}_2)|^2,$$

appropriate to the quantum mechanical state. Consistently, with this $\mathbf{X}_1$ and $\mathbf{X}_2$ are supposed to vary with time according to

$$\left.\begin{array}{l} d\mathbf{X}_1/dt = \rho(\mathbf{X}_1, \mathbf{X}_2)^{-1} \operatorname{Im} \sum_{ij} \psi_{ij}^*(\mathbf{X}_1, \mathbf{X}_2)(\partial/\partial\mathbf{X}_1)\psi_{ij}(\mathbf{X}_1, \mathbf{X}_2), \\[2ex] d\mathbf{X}_2/dt = \rho(\mathbf{X}_1, \mathbf{X}_2)^{-1} \operatorname{Im} \sum_{ij} \psi_{ij}^*(\mathbf{X}_1, \mathbf{X}_2)(\partial/\partial\mathbf{X}_2)\psi_{ij}(\mathbf{X}_1, \mathbf{X}_2). \end{array}\right\} \tag{9}$$

The curious feature is that the trajectory equations (9) for the hidden variables have in general a grossly non-local character. If the wave function is factorable before the analyzing fields become effective (the particles being far apart),

$$\psi_{ij}(\mathbf{X}_1, \mathbf{X}_2) = \phi_i(\mathbf{X}_1)\chi_j(\mathbf{X}_2),$$

this factorability will be preserved. Equation (8) then reduce to

$$d\mathbf{X}_1/dt = \left[\sum_i \phi_i^*(\mathbf{X}_1)\phi_i(\mathbf{X}_1)\right]^{-1} \operatorname{Im}\sum_i \phi_i^*(\mathbf{X}_1)(\partial/\partial\mathbf{X}_1)\phi_i(\mathbf{X}_1),$$

$$d\mathbf{X}_2/dt = \left[\sum_j \chi_j^*(\mathbf{X}_2)\chi_j(\mathbf{X}_2)\right]^{-1} \operatorname{Im}\sum_j \chi_j^*(\mathbf{X}_2)(\partial/\partial\mathbf{X}_2)\chi(\mathbf{X}_2).$$

The Schrödinger equation (8) also separates, and the trajectories of $\mathbf{X}_1$ and $\mathbf{X}_2$ are determined separately by equations involving $H(\mathbf{X}_1)$ and $H(\mathbf{X}_2)$, respectively. However, in general, the wave function is not factorable. The trajectory of 1 then depends in a complicated way on the trajectory and wave function of 2, and so on the analyzing fields acting on 2 – however remote these may be from particle 1. So in this theory an explicit causal mechanism exists whereby the disposition of one piece of apparatus affects the results obtained with a distant piece. In fact the Einstein–Podolsky–Rosen paradox is resolved in the way which Einstein would have liked least (Ref. 2, p. 85).

More generally, the hidden variable account of a given system becomes entirely different when we remember that it has undoubtedly interacted with numerous other systems in the past and that the total wave function will certainly not be factorable. The same effect complicates the hidden variable account of the theory of measurement, when it is desired to include part of the 'apparatus' in the system.

Bohm of course was well aware[6,16-18] of these features of his scheme, and has given them much attention. However, it must be stressed that, to the present writer's knowledge, there is no *proof* that *any* hidden variable account of quantum mechanics *must* have this extraordinary character.[19] It would therefore be interesting, perhaps,[1] to pursue some further 'impossibility proofs,' replacing the arbitrary axioms objected to above by some condition of locality, or of separability of distant systems.

Acknowledgements

The first ideas of this paper were conceived in 1952. I warmly thank Dr. F. Mandl for intensive discussion at that time. I am indebted to many others since then, and latterly, and very especially, to Professor J. M. Jauch.

Notes and references

1 The following works contain discussions of and references on the hidden variable problem: L. de Broglie, *Physicien et Penseur*. Albin Michel, Paris (1953); W. Heisenberg, in *Niels Bohr and the Development of Physics*, W. Pauli, Ed. McGraw-Hill Book Co., Inc., New York, and Pergamon Press, Ltd., London (1955); *Observation and Interpretation*, S. Körner, Ed. Academic Press Inc., New York, and Butterworths Scientific Publ., Ltd., London (1957); N. R. Hansen, *The Concept of the Positron*. Cambridge University Press, Cambridge, England (1963). See also the various works by D. Bohm cited later, and Bell and Nauenberg.[8] For the view that the possibility of hidden variables has little interest, see especially the contributions of Rosenfeld to the first and third of these references, of Pauli to the first, the article of Heisenberg, and many passages in Hansen.

2 A. Einstein, *Philosopher Scientist*, P. A. Schilp, Ed. Library of Living Philosophers, Evanston, Ill. (1949). Einstein's 'Autobiographical Notes' and 'Reply to Critics' suggest that the hidden variable problem has some interest.

3 J. M. Jauch and C. Piron, *Helv. Phys. Acta* **36**, 827 (1963).

4 A. M. Gleason, *J. Math. & Mech.* **6**, 885 (1957). I am much indebted to professor Jauch for drawing my attention to this work.

5 N. Bohr, in Ref. 2.

6 D. Bohm, *Phys. Rev.* **85**, 166, 180 (1952).

7 In particular the analysis of Bohm[6] seems to lack clarity, or else accuracy. He fully emphasizes the role of the experimental arrangement. However, it seems to be implied (Ref. 6, p. 187) that the circumvention of the theorem *requires* the association of *hidden* variables with the apparatus as well as with the system observed. The scheme of Section 2 is a counter example to this. Moreover, it will be seen in Section 3 that if the essential additivity assumption of von Neumann were granted, hidden variables wherever located would not avail. Bohm's further remarks in Ref. 16 (p. 95) and Ref. 17 (p. 358) are also unconvincing. Other critiques of the theorem are cited, and some of them rebutted, by Albertson (J. Albertson, *Am. J. Phys.* **29**, 478 (1961)).

8 Recent papers on the measurement process in quantum mechanics, with further references, are: E. P. Wigner, *Am. J. Phys.* **31**, 6 (1963); A. Shimony, *Am. J. Phys.* **31**, 755 (1963); J. M. Jauch, *Helv. Phys. Acta* **37**, 293 (1964); B. d'Espagnat, *Conceptions de la physique contemporaine*. Hermann & Cie., Paris (1965); J. S. Bell and M. Nauenberg, in *Preludes in Theoretical Physics, In Honor of V. Weisskopf*. North-Holland Publishing Company, Amsterdam (1966).

9 J. von Neumann, *Mathematische Grundlagen der Quanten-mechanik*. Julius Springer-Verlag, Berlin (1932) (English transl.: Princeton University Press, Princeton, N.J., 1955). All page numbers quoted are those of the English edition. The problem is posed in the preface, and on p. 209. The formal proof occupies essentially pp. 305–24 and is followed by several pages of commentary. A self-contained exposition of the proof has been presented by. J. Albertson (see Ref. 7).

10 This is contained in von Neumann's **B'** (p. 311), **1** (p. 313), and **11** (p. 314).

11 Reference 9, pp. 209.

12 Reference 9, p. 325.

13 In the two-dimensional case $\langle a \rangle = \langle b \rangle = 1$ (for some quantum mechanical state) is possible only if the two projectors are identical $(\hat{\alpha} = \hat{\beta})$. Then $a \cap b = a = b$ and $\langle a \cap b \rangle = \langle a \rangle = \langle b \rangle = 1$.

14 The simplest example for illustrating the discussion of Section 5 would then be a particle of spin 1, postulating a sufficient variety of spin–external-field interactions to permit arbitrary complete sets of spin states to be spacially separated.

15 There are clearly enough measurements to be interesting that can be made in this way. We will not consider whether there are others.

16 D. Bohm, *Causality and Chance in Modern Physics*. D. Van Nostrand Co., Inc., Princeton, N.J. (1957).

17 D. Bohm, in *Quantum Theory*, D. R. Bates, Ed. Academic Press Inc., New York (1962).

18 D. Bohm and Y. Aharonov, *Phys. Rev.* **108**, 1070 (1957).

19 Since the completion of this paper such a proof has been found (J. S. Bell, *Physics* **1**, 195 (1965)).

2

On the Einstein–Podolsky–Rosen paradox*

1 Introduction

The paradox of Einstein, Podolsky and Rosen[1] was advanced as an argument that quantum mechanics could not be a complete theory but should be supplemented by additional variables. These additional variables were to restore to the theory causality and locality[2]. In this note that idea will be formulated mathematically and shown to be incompatible with the statistical predictions of quantum mechanics. It is the requirement of locality, or more precisely that the result of a measurement on one system be unaffected by operations on a distant system with which it has interacted in the past, that creates the essential difficulty. There have been attempts[3] to show that even without such a separability or locality requirement no 'hidden variable' interpretation of quantum mechanics is possible. These attempts have been examined elsewhere[4] and found wanting. Moreover, a hidden variable interpretation of elementary quantum theory[5] has been explicitly constructed. That particular interpretation has indeed a grossly non-local structure. This is characteristic, according to the result to be proved here, of any such theory which reproduces exactly the quantum mechanical predictions.

2 Formulation

With the example advocated by Bohm and Aharonov[6], the EPR argument is the following. Consider a pair of spin one-half particles formed somehow in the singlet spin state and moving freely in opposite directions. Measurements can be made, say by Stern–Gerlach magnets, on selected components of the spins σ_1 and σ_2. If measurement of the component $\sigma_1 \cdot \mathbf{a}$, where $\mathbf{a}$ is some unit vector, yields the value $+1$ then, according to quantum mechanics, measurement of $\sigma_2 \cdot \mathbf{a}$ must yield the value -1 and vice versa. Now we make the hypothesis[2], and it seems one at least worth

* Work supported in part by the U.S. Atomic Energy Commission.
Department of Physics, University of Wisconsin, Madison, Wisconsin.

considering, that if the two measurements are made at places remote from one another the orientation of one magnet does not influence the result obtained with the other. Since we can predict in advance the result of measuring any chosen component of σ_2, by previously measuring the same component of σ_1, it follows that the result of any such measurement must actually be predetermined. Since the initial quantum mechanical wave function does not determine the result of an individual measurement, this predetermination implies the possibility of a more complete specification of the state.

Let this more complete specification be effected by means of parameters λ. It is a matter of indifference in the following whether λ denotes a single variable or a set, or even a set of functions, and whether the variables are discrete or continuous. However, we write as if λ were a single continuous parameter. The result A of measuring $\sigma_1 \cdot \mathbf{a}$ is then determined by $\mathbf{a}$ and λ, and the result B of measuring $\sigma_2 \cdot \mathbf{b}$ in the same instance is determined by $\mathbf{b}$ and λ, and

$$A(\mathbf{a}, \lambda) = \pm 1, B(\mathbf{b}, \lambda) = \pm 1. \tag{1}$$

The vital assumption[2] is that the result B for particle 2 does not depend on the setting $\mathbf{a}$, of the magnet for particle 1, nor A on $\mathbf{b}$.

If $\rho(\lambda)$ is the probability distribution of λ then the expectation value of the product of the two components $\sigma_1 \cdot \mathbf{a}$ and $\sigma_2 \cdot \mathbf{b}$ is

$$P(\mathbf{a}, \mathbf{b}) = \int d\lambda \rho(\lambda) A(\mathbf{a}, \lambda) B(\mathbf{b}, \lambda) \tag{2}$$

This should equal the quantum mechanical expectation value, which for the singlet state is

$$\langle \sigma_1 \cdot \mathbf{a}\, \sigma_2 \cdot \mathbf{b} \rangle = -\,\mathbf{a} \cdot \mathbf{b} \tag{3}$$

But it will be shown that this is not possible.

Some might prefer a formulation in which the hidden variables fall into two sets, with A dependent on one and B on the other; this possibility is contained in the above, since λ stands for any number of variables and the dependences thereon of A and B are unrestricted. In a complete physical theory of the type envisaged by Einstein, the hidden variables would have dynamical significance and laws of motion; our λ can then be thought of as initial values of these variables at some suitable instant.

3 Illustration

The proof of the main result is quite simple. Before giving it, however, a number of illustrations may serve to put it in perspective.

Firstly, there is no difficulty in giving a hidden variable account of spin measurements on a single particle. Suppose we have a spin half particle in a pure spin state with polarization denoted by a unit vector **p**. Let the hidden variable be (for example) a unit vector λ with uniform probability distribution over the hemisphere $\lambda \cdot \mathbf{p} > 0$. Specify that the result of measurement of a component $\boldsymbol{\sigma} \cdot \mathbf{a}$ is

$$\text{sign } \lambda \cdot \mathbf{a}', \tag{4}$$

where $\mathbf{a}'$ is a unit vector depending on **a** and **p** in a way to be specified, and the sign function is $+1$ or -1 according to the sign of its argument. Actually this leaves the result undetermined when $\lambda \cdot \mathbf{a}' = 0$, but as the probability of this is zero we will not make special prescriptions for it. Averaging over λ the expectation value is

$$\langle \boldsymbol{\sigma} \cdot \mathbf{a} \rangle = 1 - 2\theta'/\pi, \tag{5}$$

where θ' is the angle between $\mathbf{a}'$ and **p**. Suppose then that $\mathbf{a}'$ is obtained from **a** by rotation towards **p** until

$$1 - \frac{2\theta'}{\pi} = \cos\theta \tag{6}$$

where θ is the angle between **a** and **p**. Then we have the desired result

$$\langle \boldsymbol{\sigma} \cdot \mathbf{a} \rangle = \cos\theta \tag{7}$$

So in this simple case there is no difficulty in the view that the result of every measurement is determined by the value of an extra variable, and that the statistical features of quantum mechanics arise because the value of this variable is unknown in individual instances.

Secondly, there is no difficulty in reproducing, in the form (2), the only features of (3) commonly used in verbal discussions of this problem:

$$\left.\begin{array}{l} P(\mathbf{a}, \mathbf{a}) = -P(\mathbf{a}, -\mathbf{a}) = -1 \\ P(\mathbf{a}, \mathbf{b}) = 0 \quad \text{if } \mathbf{a} \cdot \mathbf{b} = 0 \end{array}\right\} \tag{8}$$

For example, let λ now be unit vector λ, with uniform probability distribution over all directions, and take

$$\left.\begin{array}{l} A(\mathbf{a}, \lambda) = \text{sign } \mathbf{a} \cdot \lambda \\ B(a, b) = -\text{sign } \mathbf{b} \cdot \lambda \end{array}\right\} \tag{9}$$

This gives

$$P(\mathbf{a}, \mathbf{b}) = -1 + \frac{2}{\pi}\theta, \tag{10}$$

where θ is the angle between a and b, and (10) has the properties (8). For comparison, consider the result of a modified theory[6] in which the pure singlet state is replaced in the course of time by an isotropic mixture of product states; this gives the correlation function

$$-\frac{1}{3}\mathbf{a}\cdot\mathbf{b} \tag{11}$$

It is probably less easy, experimentally, to distinguish (10) from (3), than (11) from (3).

Unlike (3), the function (10) is not stationary at the minimum value -1 (at $\theta = 0$). It will be seen that this is characteristic of functions of type (2).

Thirdly, and finally, there is no difficulty in reproducing the quantum mechanical correlation (3) if the results A and B in (2) are allowed to depend on $\mathbf{b}$ and $\mathbf{a}$ respectively as well as on $\mathbf{a}$ and $\mathbf{b}$. For example, replace $\mathbf{a}$ in (9) by $\mathbf{a}'$, obtained from $\mathbf{a}$ by rotation towards $\mathbf{b}$ until

$$1 - \frac{2}{\pi}\theta' = \cos\theta,$$

where θ' is the angle between $\mathbf{a}'$ and $\mathbf{b}$. However, for given values of the hidden variables, the results of measurements with one magnet now depend on the setting of the distant magnet, which is just what we would wish to avoid.

4 Contradiction

The main result will now be proved. Because ρ is a normalized probability distribution,

$$\int d\lambda \rho(\lambda) = 1, \tag{12}$$

and because of the properties (1), P in (2) cannot be less than -1. It can reach -1 at $\mathbf{a} = \mathbf{b}$ only if

$$A(\mathbf{a}, \lambda) = -B(\mathbf{a}, \lambda) \tag{13}$$

except at a set of points λ of zero probability. Assuming this, (2) can be rewritten

$$P(\mathbf{a}, \mathbf{b}) = -\int d\lambda \rho(\lambda) A(\mathbf{a}, \lambda) A(\mathbf{b}, \lambda). \tag{14}$$

It follows that if **c** is another unit vector

$$P(\mathbf{a}, \mathbf{b}) - P(\mathbf{a}, \mathbf{c}) = - \int d\lambda \rho(\lambda) [A(\mathbf{a}, \lambda) A(\mathbf{b}, \lambda) - A(\mathbf{a}, \lambda) A(\mathbf{c}, \lambda)]$$

$$= \int d\lambda \rho(\lambda) A(\mathbf{a}, \lambda) A(\mathbf{b}, \lambda) [A(\mathbf{b}, \lambda) A(\mathbf{c}, \lambda) - 1]$$

using (1), whence

$$|P(\mathbf{a}, \mathbf{b}) - P(\mathbf{a}, \mathbf{c})| \leqslant \int d\lambda \rho(\lambda) [1 - A(\mathbf{b}, \lambda) A(\mathbf{c}, \lambda)]$$

The second term on the right is $P(\mathbf{b}, \mathbf{c})$, whence

$$1 + P(\mathbf{b}, \mathbf{c}) \geqslant |P(\mathbf{a}, \mathbf{b}) - P(\mathbf{a}, \mathbf{c})| \tag{15}$$

Unless P is constant, the right hand side is in general of order $|\mathbf{b} - \mathbf{c}|$ for small $|\mathbf{b} - \mathbf{c}|$. Thus $P(\mathbf{b}, \mathbf{c})$ cannot be stationary at the minimum value (-1 at $\mathbf{b} = \mathbf{c}$) and cannot equal the quantum mechanical value (3).

Nor can the quantum mechanical correlation (3) be arbitrarily closely approximated by the form (2). The formal proof of this may be set out as follows. We would not worry about failure of the approximation at isolated points, so let us consider instead of (2) and (3) the functions

$$\bar{P}(\mathbf{a}, \mathbf{b}) \quad \text{and} \quad \overline{- \mathbf{a} \cdot \mathbf{b}}$$

where the bar denotes independent averaging of $P(\mathbf{a}', \mathbf{b}')$ and $- \mathbf{a}' \cdot \mathbf{b}'$ over vectors $\mathbf{a}'$ and $\mathbf{b}'$ within specified small angles of $\mathbf{a}$ and $\mathbf{b}$. Suppose that for all $\mathbf{a}$ and $\mathbf{b}$ the difference is bounded by ε:

$$|\bar{P}(\mathbf{a}, \mathbf{b}) + \mathbf{a} \cdot \mathbf{b}| \leqslant \varepsilon \tag{16}$$

Then it will be shown that ε cannot be made arbitrarily small.

Suppose that for all a and b

$$|\overline{\mathbf{a} \cdot \mathbf{b}} - \mathbf{a} \cdot \mathbf{b}| \leqslant \delta \tag{17}$$

Then from (16)

$$|\bar{P}(\mathbf{a}, \mathbf{b}) + \mathbf{a} \cdot \mathbf{b}| \leqslant \varepsilon + \delta \tag{18}$$

From (2)

$$\bar{P}(\mathbf{a}, \mathbf{b}) = \int d\lambda \rho(\lambda) \bar{A}(\mathbf{a}, \lambda) \bar{B}(\mathbf{b}, \lambda) \tag{19}$$

where

$$|\bar{A}(\mathbf{a}, \lambda)| \leqslant 1 \quad \text{and} \quad |\bar{B}(\mathbf{b}, \lambda)| \leqslant 1 \qquad (20)$$

From (18) and (19), with $\mathbf{a} = \mathbf{b}$,

$$d\lambda\rho(\lambda)[\bar{A}(\mathbf{b}, \lambda)\bar{B}(\mathbf{b}, \lambda) + 1] \leqslant \varepsilon + \delta \qquad (21)$$

From (19)

$$\bar{P}(\mathbf{a}, \mathbf{b}) - \bar{P}(\mathbf{a}, \mathbf{c}) = \int d\lambda\rho(\lambda)[\bar{A}(\mathbf{a}, \lambda)\bar{B}(\mathbf{b}, \lambda) - \bar{A}(\mathbf{a}, \lambda)\bar{B}(\mathbf{c}, \lambda)]$$

$$= \int d\lambda\rho(\lambda)\bar{A}(\mathbf{a}, \lambda)\bar{B}(\mathbf{b}, \lambda)[1 + \bar{A}(\mathbf{b}, \lambda)\bar{B}(\mathbf{c}, \lambda)]$$

$$- \int d\lambda\rho(\lambda)\bar{A}(\mathbf{a}, \lambda)\bar{B}(\mathbf{c}, \lambda)[1 + \bar{A}(\mathbf{b}, \lambda)\bar{B}(\mathbf{b}, \lambda)]$$

Using (20) then

$$|\bar{P}(\mathbf{a}, \mathbf{b}) - \bar{P}(\mathbf{a}, \mathbf{c})| \leqslant \int d\lambda\rho(\lambda)[1 + \bar{A}(\mathbf{b}, \lambda)\bar{B}(\mathbf{c}, \lambda)]$$

$$+ \int d\lambda\rho(\lambda)[1 + \bar{A}(\mathbf{b}, \lambda)\bar{B}(\mathbf{b}, \lambda)]$$

Then using (19) and (21)

$$|\bar{P}(\mathbf{a}, \mathbf{b}) - \bar{P}(\mathbf{a}, \mathbf{c})| \leqslant 1 + \bar{P}(\mathbf{b}, \mathbf{c}) + \varepsilon + \delta$$

Finally, using (18),

$$|\mathbf{a} \cdot \mathbf{c} - \mathbf{a} \cdot \mathbf{b}| - 2(\varepsilon + \delta) \leqslant 1 - \mathbf{b} \cdot \mathbf{c} + 2(\varepsilon + \delta)$$

or

$$4(\varepsilon + \delta) \geqslant |\mathbf{a} \cdot \mathbf{c} - \mathbf{a} \cdot \mathbf{b}| + \mathbf{b} \cdot \mathbf{c} - 1 \qquad (22)$$

Take for example $\mathbf{a} \cdot \mathbf{c} = 0$, $\mathbf{a} \cdot \mathbf{b} = \mathbf{b} \cdot \mathbf{c} = 1/\sqrt{2}$. Then

$$4(\varepsilon + \delta) \geqslant \sqrt{2} - 1$$

Therefore, for small finite δ, ε cannot be arbitrarily small.

Thus, the quantum mechanical expectation value cannot be represented, either accurately or arbitrarily closely, in the form (2).

5 Generalization

The example considered above has the advantage that it requires little imagination to envisage the measurements involved actually being made.

In a more formal way, assuming[7] that any Hermitian operator with a complete set of eigenstates is an 'observable', the result is easily extended to other systems. If the two systems have state spaces of dimensionality greater than 2 we can always consider two-dimensional subspaces and define, in their direct product, operators σ_1 and σ_2 formally analogous to those used above and which are zero for states outside the product subspace. Then for at least one quantum mechanical state, the 'singlet' state in the combined subspaces, the statistical predictions of quantum mechanics are incompatible with separable predetermination.

6 Conclusion

In a theory in which parameters are added to quantum mechanics to determine the results of individual measurements, without changing the statistical predictions, there must be a mechanism whereby the setting of one measuring device can influence the reading of another instrument, however remote. Moreover, the signal involved must propagate instantaneously, so that such a theory could not be Lorentz invariant.

Of course, the situation is different if the quantum mechanical predictions are of limited validity. Conceivably they might apply only to experiments in which the settings of the instruments are made sufficiently in advance to allow them to reach some mutual rapport by exchange of signals with velocity less than or equal to that of light. In that connection, experiments of the type proposed by Bohm and Aharonov[6], in which the settings are changed during the flight of the particles, are crucial.

Acknowledgement

I am indebted to Drs. M. Bander and J. K. Perring for very useful discussions of this problem. The first draft of the paper was written during a stay at Brandeis University; I am indebted to colleagues there and at the University of Wisconsin for their interest and hospitality.

Notes and references

1   A. Einstein, N. Rosen and B. Podolsky, *Phys. Rev.* **47**, 777 (1935); see also N. Bohr, *Phys. Rev.* **48**, 696 (1935), W. H. Furry, *Phys. Rev.* **49**, 393 and 476 (1936), and D. R. Inglis, *Rev. Mod. Phys.* **33**, 1(1961).

2   'But on one supposition we should, in my opinion, absolutely hold fast: the real factual situation of the system S_2 is independent of what is done with the system S_1, which is spatially separated from the former.' A. Einstein in *Albert Einstein, Philosopher Scientist*, Edited by P. A. Schilp, p. 85, Library of Living Philosophers, Evanston, Illinois (1949).

3 J. von Neumann, *Mathematische Grundlagen der Quanten-mechanik.* Verlag Julius-Springer, Berlin (1932), (English translation: Princeton University Press (1955); J. M. Jauch and C. Piron, *Helv. Phys. Acta* **36**, 827 (1963).

4 J. S. Bell, *Rev. Mod. Phys.* **38**, 447 (1966).

5 D. Bohm, *Phys. Rev.* **85**, 166 and 180 (1952).

6 D. Bohm and Y. Aharonov, *Phys. Rev.* **108**, 1070 (1957).

7 P. A. M. Dirac, *The Principles of Quantum Mechanics* (3rd Ed.) p. 37. The Clarendon Press, Oxford (1947).

3

*The moral aspect of quantum mechanics**

The notion of morality appears to have been introduced into quantum theory by Wigner, as reported by Goldberger and Watson[1]. The question at issue is the famous 'reduction of the wave packet'. There are, ultimately, no mechanical arguments for this process, and the arguments that are actually used may well be called moral. This is a popular account of the subject. Very practical people not interested in logical questions should not read it. It is a pleasure for us to dedicate the paper to Professor Weisskopf, for whom intense interest in the latest developments of detail has not dulled concern with fundamentals.

Suppose that some quantity F is measured on a quantum mechanical system, and a result f obtained. Assume that immediate repetition of the measurement must give the same result. Then, after the first measurement, the system must be in an eigenstate of F with eigenvalue f. In general, the measurement will be 'incomplete', i.e., there will be more than one eigenstate with the observed eigenvalue, so that the latter does not suffice to specify completely the state resulting from the measurement. Let the relevant set of eigenstates be donoted by ϕ_{fg}. The extra index g may be regarded as the eigenvalue of a second observable G that commutes with F and so can be measured at the same time. Given that f is observed for F, the relative probabilities of observing various g in a simultaneous measurement of G are given by the squares of the moduli of the inner products

$$(\phi_{fg}, \psi)$$

where ψ is the initial state of the system. Let us now make the plausible assumption that these relative probabilities would be the same if G were measured not simultaneously with F but immediately afterwards. Then we know something more about the state resulting from the measurement of F. One state with the desired properties is clearly

$$N \sum_g \phi_{fg}(\phi_{fg}, \psi)$$

* With M. Nauenberg, Stanford University.

where N is a normalization factor. It is readily shown that this is the only state[2] for which the probability of obtaining a given value for *any* quantity commuting with F is the same whether the measurement is made at the same time or immediately after. Thus, we arrive at the general formulation for the 'reduction of the wave packet' following measurement[3]: expand the initial state in eigenstates of the observed quantity, strike out the contributions from eigenstates which do not have the observed eigenvalue, and renormalize the remainder. This preserves the original phase and intensity relations between the relevant eigenstates. It therefore does the minimum damage to the original state consistent with the requirement that an immediate repetition of the measurement gives the same result. All this is very ethical, and we will refer to the particular reduction just defined as 'the moral process'.

Now morality is not universally observed, and it is easy to think of measuring processes for which the above account would be quite inappropriate. Suppose for example the momentum of a neutron is measured by observing a recoil proton. The momentum of the neutron is altered in the process, and in a head-on collision actually reduced to zero. The subsequent state of the neutron is by no means a combination (the spin here provides the degeneracy) of states with the observed momentum. How then is one to know whether a given measurement is moral[4] or not? Clearly, one must investigate the physics of the process. Instead of tracing through a realistic example we will follow von Neumann[3] here in considering a simple model.

Suppose the system I to be observed has coordinates R. Suppose that the measuring instrument, II, has a single relevant coordinate Q – a pointer position. Suppose that the measurement is effected by switching on instantaneously an interaction between I and II

$$\delta(t)F\left(R, \frac{1}{i}\frac{\partial}{\partial R}\right)\frac{1}{i}\frac{\partial}{\partial Q}$$

where t is time. The simplification here, where the system of interest acts directly on a pointer reading without intervention of circuitry, is gross. If I is in the state $\psi(R)$ before the measurement, and the pointer reading is zero, the initial state of I + II is

$$\psi(R)\delta(Q).$$

The state of I + II immediately after $t = 0$ can be obtained by solving the Schrödinger equation. In this only the interaction term in the Hamiltonian is significant, because of its impulsive character. The resulting state is[5]

$$\sum_{f,g} \phi_{fg}(R)(\phi_{fg}, \psi)\delta(Q - f)$$

where f is an eigenvalue of F, ϕ_{fg} a corresponding eigenfunction, and g any extra index needed to enumerate these eigenfunctions. If now an observer reads the pointer on the instrument, and finds a particular value f, *and if this measurement of the pointer reading is moral*, then the state reduces to

$$N \sum_g \phi_{fg}(R)(\phi_{fg}, \psi)\delta(Q - f).$$

The part referring to system I alone,

$$N \sum_g \phi_{fg}(R)(\phi_{fg}, \psi)$$

is precisely the result of applying the moral process to I directly, after the measurement of the quantity F. So we have here a dynamical model of a moral measurement of F. This depends on the detailed nature of the interaction between the system and the measuring instrument. It would have been equally easy to choose an interaction for which a moral measurement of the pointer reading would imply an immoral measurement of F.

Thus, if the morality of measurements of macroscopic pointer readings is granted, there is no real ambiguity in practice in applying quantum mechanics. One must simply understand well enough the structure of the systems involved, including the instruments, and work out the consequences. This situation is not peculiar to quantum mechanics. Moreover, we are readily disposed to accept the moral character of observing macroscopic pointers, for we feel convinced from common experience that they are not much changed in state by being looked at, and the moral process is in an obvious sense minimal. Thus, the basis of practical quantum mechanics seems secure. This is just as well, in view of its magnificent success, and of the fact that there is no real competitor in sight. However, it must not be supposed that the action on the wave function of even such a macroscopic observation is of a trivial nature, and least of all that it is a mere subjective adjustment of the representative ensemble to allow for increased knowledge. To make this elementary point suppose that the measuring interaction in the above model is again switched on at times τ and 2τ:

$$\delta(t - \tau)F\frac{1}{i}\frac{\partial}{\partial Q}, \quad \delta(t - 2\tau)F\frac{1}{i}\frac{\partial}{\partial Q}.$$

During the period τ suppose that each eigenstate ϕ_f (the possible extra index g is not essential here) evolves into a combination

$$\phi_f \rightarrow \sum_{f'} \phi_{f'}\alpha_{f',f}.$$

For the instrument II suppose for simplicity that Q is a constant of the motion between interactions. Then solution of the Schrödinger equation for I + II gives from the initial state (just before $t = 0$)

$$\psi\delta(Q)$$

the final state

$$\sum_{f,f',f''} \phi_{f''}\alpha_{f'',f'}\alpha_{f',f}(\phi_f,\psi)\delta(Q-f-f'-f'')$$

just after $t = 2\tau$. The probabilities of then observing various particular possible values Q for the pointer position are given by

$$\sum_{f''}\left|\sum_f \alpha_{f'',Q-f-f''}\alpha_{Q-f-f'',f}(\phi_f,\psi)\right|^2.$$

Now this assumes that the intermediate evolution of I + II is governed entirely by the Schrödinger equation, and therefore *that the pointer position is not looked at until after the final interaction*. If the pointer position is observed just after *each* interaction then the moral process comes into play just after $t = 0$ and $t = \tau$. If all possible results of these intermediate observations are averaged over, the net result is simply to eliminate from the last expression interference between different values of f and f'; it becomes

$$\sum_{f''}\sum_f |\alpha_{f'',Q-f-f''}\alpha_{Q-f-f'',f}(\phi_f,\psi)|^2.$$

Thus observation, even when all possible results are averaged over, is a dynamical interference with the system which may alter the statistics of subsequent measurements.

Now although we would not wish to cast doubt on the *practical* adequacy of macroscopic morality, it is clear that if we leave it un-analyzed the theory can at best be described as a phenomenological makeshift. The fact already stressed that observation implies a dynamical interference, together with the belief that instruments after all are no more than large assemblies of atoms, and that they interact with the rest of the world largely through the well-known electromagnetic interaction, seems to make this a distinctly uncomfortable level at which to replace analysis by axioms. The only possibility of further analysis offered by quantum mechanics is to incorporate still more of the world into the quantum mechanical system, I + II + III + etc. Especially from the theorist's point of view such a development is very pertinent. For him the experiment may be said to start with the printed proposal and to end with the issue of the report. For him the laboratory, the experimenter, the administration, and the editorial staff of the *Physical Review*, are all just part of the instrumentation. The incorporation of (presumably) conscious experimenters and editors into the equipment raises a very intriguing question. For they know the results before the theorist reads the report, and the question is whether their knowledge is incompatible with the sort of interference phenomena discussed above. If the interference is destroyed, then the Schrödinger equation is incorrect for systems containing consciousness. If the interference is not destroyed the quantum mechanical description is revealed as not wrong but certainly incomplete[8]. We have something analogous to a two-slit interference experiment where the '*particle*' in any particular instance has gone through only one of the slits (and knows it!) and yet there are interference terms depending on the *wave* having gone through both slits. Thus we have *both* waves *and* particle trajectories, as in the de Broglie–Bohm 'pilot wave' or 'hidden parameter' interpretations of quantum mechanics[7]. Unfortunately it seems hopelessly impossible to test this question in practice; it is hard enough to realize interference phenomena involving simple things like electrons, photons, or α particles. Experimenters (and even inanimate instruments) radiate heat, for example, and this coupling to their surroundings suppresses interference just as effectively as the theorist reading the *Physical Review*. Nevertheless, the question of principle is there. Now, even if we had settled the status of the experimenter, we are not at the end of the road. For the reading of the *Physical Review* is hardly a more elementary act than the reading of pointers or computer output; this act also seems to require analysis rather than axiomatics, and so we want the theorist also in the Schrödinger equation. He also radiates heat, and so on, and we want finally the whole universe in the quantum mechanical system. At this point we are finally lost. It is easy to imagine a state vector for the whole universe, quietly pursuing its linear evolution through all of time and containing somehow all possible worlds. But the usual interpretive axioms of quantum mechanics come into play only when the system interacts with something else, is 'observed'. For the universe there *is* nothing else, and quantum mechanics in its traditional form has simply nothing to say. It gives no way of, indeed no meaning in, picking out from the wave of possibility the single unique thread of history.

These considerations, in our opinion, lead inescapably to the conclusion that quantum mechanics is, at the best, incomplete[8]. We look forward to a new theory which can refer meaningfully to events in a given system without requiring 'observation' by another system. The critical test cases requiring this conclusion are systems containing consciousness and the universe as a whole. Actually, the writers share with most physicists a degree of embarrassment at consciousness being dragged into physics, and share the usual feeling that to consider the universe as a whole is at least immodest, if not blasphemous. However, these are only logical test cases. It seems likely to us that physics will have again adopted a more objective description of nature long before it begins to understand consciousness, and the universe as a whole may well play no central role in this development. It remains a logical possibility that it is the act of consciousness which is ultimately responsible for the reduction of the wave packet[9]. It is also possible that something like the quantum mechanical state function continue to play a role, supplemented by variables describing the actual as distinct from the possible course of events ('hidden variables') although this approach seems to face severe difficulties in describing separated systems in a sensible way[7]. What is much more likely is that the new way of seeing things will involve an imaginative leap that will astonish us. In any case it seems that the quantum mechanical description will be superseded. In this it is like all theories made by man. But to an unusual extent its ultimate fate is apparent in its internal structure. It carries in itself the seeds of its own destruction.

Notes and references

1 M. L. Goldberger and K. M. Watson, *Phys. Rev.* **134**, B919 (1964).

2 To show formally that there is no other such state it suffices to consider as second observable the projection operator on to an arbitrary combination of states ϕ_{fg} with the given f. The set of expectation values of all such projections determines the state.

3 J. von Neumann, Mathematische Grundlagen der Quantenmechanik, Verlag Julius Springer, Berlin (1932) (Eng. trans. Princeton Univ. Press, 1955) Chapter 6. The prescription for incomplete measurement is implicit is most treatments of quantum measurement theory, for example that of von Neumann. It is not often stated explicitly. See, however, F. Mandl, Quantum Mechanics, 2nd edition. Butterworth, London, p. 69 (1957), and the references to A. Messiah and E. P. Wigner cited by Goldberger and Watson in Ref. 1.

4 Moral and immoral measurements were called respectively measurements of the first and second kind by W. Pauli in Handbuch der Physik, Vol. V/1, p. 72. Springer-Verlag, Berlin (1957).

5 This can be obtained by noting that the state

$$\chi = \phi_{fg}\delta(Q - \alpha(t)f)$$

satisfies

$$\frac{\partial \chi}{\partial t} = -\frac{d\alpha}{dt} f \frac{\partial \chi}{\partial Q} = -i\frac{d\alpha}{dt} F \frac{1}{i} \frac{\partial \chi}{\partial Q}.$$

So we need $(d\alpha/dt) = \delta(t)$, or that α increases from zero to one during the interaction. Given in the text is the combination of such solutions which corresponds to the prescribed initial state.

6 It is taken for granted here that conscious experience is of, or is, a unique sequence of events, and cannot be completely described by a quantum mechanical state containing somehow all possible sequences. Occasionally people challenge this view. The writers therefore concede that there may be *some* people whose states of mind are best described by coherent or incoherent quantum mechanical superpositions.

7 For references on this approach and analysis of some objections to it see J. S. Bell, *Rev. Mod. Phys.*, Oct. 1965. For a more serious objection see J. S. Bell, *Physics* 1, 195 (1965).

8 This minority view is as old as quantum mechanics itself, so the new theory may be a long time coming. For a recent expression of the view that on the contrary there is no real problem, only a 'pseudoproblem', see J. M. Jauch, *Helv. Phys. Acta* 37, 293 (1964). The references in that paper, and in the papers of Ref. (7), allow much of the extensive literature to be traced. We emphasize not only that our view is that of a minority, but also that current interest in such questions is small. The typical physicist feels that they have long been answered, and that he will fully understand just how if ever he can spare twenty minutes to think about it.

9 See, for example, F. London and E. Bauer, Théorie de l'observation en méchanique quantique. Hermann, Paris (1939) p. 41, or more recently E. P. Wigner in The Scientist Speculates. R. Good, Ed., Heinemann, London (1962).

4

Introduction to the hidden-variable question

1 Motivation

Theoretical physicists live in a classical world, looking out into a quantum-mechanical world. The latter we describe only subjectively, in terms of procedures and results in our classical domain. This subjective description is effected by means of quantum-mechanical state functions ψ, which characterize the classical conditioning of quantum-mechanical systems and permit predictions about subsequent events at the classical level. The classical world of course is described quite directly – 'as it is'. We could specify for example the actual positions $\Lambda_1, \Lambda_2, \cdots$ of material bodies, such as the switches defining experimental conditions and the pointers, or print, defining experimental results. Thus in contemporary theory the most complete description of the state of the world as a whole, or of any part of it extending into our classical domain, is of the form

$$(\Lambda_1, \Lambda_2, \ldots, \psi) \tag{1}$$

with both classical variables and one or more quantum-mechanical wave functions.

Now nobody knows just where the boundary between the classical and quantum domain is situated. Most feel that experimental switch settings and pointer readings are on this side. But some would think the boundary nearer, others would think it farther, any many would prefer *not* to think about it. In fact, the matter is of very little importance in practice. This is because of the immense difference in scale between things for which quantum-mechanical description is numerically essential and those ordinarily perceptible by human beings. Nevertheless, the movability of the boundary is of only approximate validity; demonstrations of it depend on neglecting numbers which are small, but not zero, which might tend to zero for infinitely large systems, but are only very small for real finite systems. A theory founded in this way on arguments of manifestly approximate character, however good the approximation, is surely of provisional nature.

It seems legitimate to speculate on how the theory might evolve. But of course no one is obliged to join in such speculation.

A possibility is that we find exactly where the boundary lies. More plausible to me is that we will find that there is no boundary. It is hard for me to envisage intelligible discourse about a world with no classical part – no base of given events, be they only mental events in a single consciousness, to be correlated. On the other hand, it is easy to imagine that the classical domain could be extended to cover the whole. The wave functions would prove to be a provisional or incomplete description of the quantum-mechanical part, of which an objective account would become possible. It is this possibility, of a homogeneous account of the world, which is for me the chief motivation of the study of the so-called 'hidden variable' possibility.

A second motivation is connected with the statistical character of quantum-mechanical predictions. Once the incompleteness of the wave-function description is suspected, it can be conjectured that the seemingly random statistical fluctuations are determined by the extra 'hidden' variables – 'hidden' because at this stage we can only conjecture their existence and certainly cannot control them. Analogously, the description of Brownian motion for example might first have been developed in a purely statistical way, the statistics becoming intelligible later with the hypothesis of the molecular constitution of fluids, this hypothesis then pointing to previously unimagined experimental possibilities, the exploitation of which made the hypothesis entirely convincing. For me the possibility of determinism is less compelling than the possibility of having one world instead of two. But, by requiring it, the programme becomes much better defined and more easy to come to grips with.

A third motivation is in the peculiar character of some quantum-mechanical predictions, which seem almost to cry out for a hidden variable interpretation. This is the famous argument of Einstein, Podolsky and Rosen[1]. Consider the example, advanced by Bohm[2], of a pair of spin-$\frac{1}{2}$ particles formed somehow in the singlet spin state and then moving freely in opposite directions. Measurements can be made, say by Stern–Gerlach magnets, on selected components of the spins σ_1 and σ_2. If measurement of $\sigma_1 \cdot \mathbf{a}$, where $\mathbf{a}$ is some unit vector, yields the value $+1$, then, according to quantum mechanics, measurement of $\sigma_2 \cdot \mathbf{a}$ must yield the value -1, and *vice versa*. Thus we can know in advance the result of measuring any component of σ_2 by previously, and possibly at a very distant place, measuring the corresponding component of σ_1. This strongly suggests that the outcomes of such measurements, along arbitrary directions, are actually determined in advance, by variables over which we have no control, but which are sufficiently revealed by the first measurement so that we can anticipate the result of the second. There need then be no temptation to regard the performance of one measurement as a causal influence on the result of the second, distant, measurement. The description of the situation could be manifestly 'local'. This idea seems at least to merit investigation.

We will find, in fact, that no local deterministic hidden-variable theory can reproduce all the experimental predictions of quantum mechanics. This opens the possibility of bringing the question into the experimental domain, by trying to approximate as well as possible the idealized situations in which local hidden variables and quantum mechanics cannot agree. However, before coming to this, we must clear the ground by some remarks on various mathematical investigations that have been made on the possibility of hidden variables in quantum mechanics without any reference to locality.

2 The absence of dispersion-free states in various formalisms derived from quantum mechanics

Consider first the usual Heisenberg uncertainty principle. It says that for quantum-mechanical states the predictions for measurements for at least one of a pair of conjugate variables must be statistically uncertain. Thus no quantum-mechanical state can be 'dispersion-free' for every observable. It follows that if a hidden-variable account is possible, in which the results of all observations are fully determined, each quantum-mechanical state must correspond to an ensemble of states each with different values of the hidden variables. Only these component states will be dispersion-free. So one way to formulate the hidden-variable problem is a search for a formalism permitting such dispersion-free states.

An early, and very celebrated, example of such an investigation was that of von Neumann[3]. He observed that in quantum mechanics an observable whose operator is a linear combination of operators for other observables

$$A = \beta B + \gamma C$$

has for expectation value the corresponding linear combination of expectation values:

$$\langle A \rangle = \beta \langle B \rangle + \gamma \langle C \rangle. \tag{2}$$

He considered more general schemes in which this particular feature was preserved. Now for the hypothetical dispersion-free states there is no distinction between expectation values and eigenvalues – for each such state must yield with certainty a particular one of the possible results for any measurement. But eigenvalues are not additive. Consider for example components of spin for a particle of spin $\frac{1}{2}$. The operator for the component along the direction half-way between the x and y axes is

$$(\sigma_x + \sigma_y)/\sqrt{2},$$

whose eigenvalues ± 1 are certainly not the corresponding linear combinations

$$(\pm 1 \pm 1)/\sqrt{2}$$

of eigenvalues of σ_x and σ_y. Thus the requirement of additive expectation values excludes the possibility of dispersion-free states. Von Neumann concluded that a hidden-variable interpretation is not possible for quantum mechanics: 'it is therefore not, as is often assumed, a question of re-interpretation of quantum mechanics – the present system of quantum mechanics would have to be objectively false in order that another description of the elementary process than the statistical one be possible'.

It seems therefore that von Neumann considered the additivity (2) more as an obvious axiom than as a possible postulate. But consider what it means in terms of the actual physical situation. Measurements of the three quantities

$$\sigma_x, \quad \sigma_y, \quad (\sigma_x + \sigma_y)/\sqrt{2},$$

require three different orientations of the Stern–Gerlach magnet, and cannot be performed simultaneously. It is just this which makes intelligible the non-additivity of the eigenvalues – the values observed in specific instances. It is by no means a question of simply measuring different components of a pre-existing vector, but rather of observing different products of different physical procedures. That the statistical averages should then turn out to be additive is really a quite remarkable feature of quantum-mechanical states, which could not be guessed *a priori*. It is by no means a 'law of thought' and there is no *a priori* reason to exclude the possibility of states for which it is false. It can be objected that although the additivity of expectation values is not a law of thought, it *is* after all experimentally true. Yes, but what we are now investigating is precisely the hypothesis that the states presented to us by nature are in fact mixtures of component states which we cannot (for the present) prepare individually. The component states need only have such properties that ensembles of them have the statistical properties of observed states.

It has subsequently been shown that in various other mathematical schemes, derived from quantum mechanics, dispersion-free states are not possible[4]. The persistence in these schemes of a kind of uncertainty principle is of course useful and interesting to people working with those schemes. However, the importance of these results, for the question that we are concerned with, is easily exaggerated. The postulates often have great intrinsic appeal to those approaching quantum mechanics in an abstract way. Translated into assumptions about the behaviour of actual physical equipment, they are again seen to be of a far from trivial or inevitable nature[4].

On the other hand, if no restrictions whatever are imposed on the hidden variables, or on the dispersion-free states, it is trivially clear that such schemes can be found to account for any experimental results whatever. Ad hoc schemes of this kind are devised every day when experimental physicists, to optimize the design of their equipment, simulate the expected results by deterministic computer programmes drawing on a table of random numbers. Such schemes, from our present point of view, are not very interesting. Certainly what Einstein wanted was a comprehensive account of physical processes evolving continuously and locally in ordinary space and time. We proceed now to describe a very instructive attempt in that direction.

3 A simple example

Consider the simple hidden-variable picture of elementary wave mechanics advanced originally by de Broglie[5] and subsequently clarified by Bohm[6]. Take the case of a single particle of spin $\frac{1}{2}$ moving in a magnetic field $\mathbf{H}$. The Schrödinger equation is

$$i\frac{\partial}{\partial t}\psi(\mathbf{r},t) = \left\{\frac{1}{2m}\left(\frac{1}{i}\frac{\partial}{\partial\mathbf{r}}\right)^2 + \mu\boldsymbol{\sigma}\cdot\mathbf{H}\right\}\psi(\mathbf{r},t), \tag{3}$$

where the wave function ψ is a two-component Pauli spinor. Let us supplement this quantum-mechanical picture by an additional (hidden) variable λ, a single three-vector, which evolves as a function of time according to the law

$$\frac{d\lambda}{dt} = \frac{\mathbf{j}_\psi(\lambda,t)}{\varrho_\psi(\lambda,t)}, \tag{4}$$

where $\mathbf{j}$ and ϱ are probability currents and densities calculated in the usual way

$$\mathbf{j}_\psi(\mathbf{r},t) = \frac{1}{2}\mathrm{Im}\psi^*(\mathbf{r},t)\frac{\partial}{\partial\mathbf{r}}\psi(\mathbf{r},t),$$

$$\varrho_\psi(\mathbf{r},t) = \psi^*(\mathbf{r},t)\psi(\mathbf{r},t),$$

with summation over suppressed spinor indices understood. It is supposed
that the quantum-mechanical state specified by the wave function ψ
corresponds to an ensemble of states (λ, ψ) in which the λs occur with
probability density $\varrho(\lambda, t)$ such that

$$\varrho(\lambda, t) = \varrho_\psi(\lambda, t).$$

It is easy to see that if the distribution ϱ of λ is equal to ϱ_ψ in this way at some
initial time, then in virtue of the equations of motion (3) and (4) it remains so
at later times.

The fundamental interpretative rule of the model is just that $\lambda(t)$ is the
real position of the particle at time t, and that observation of position will
yield this value. Thus the quantum statistics of position measurements, the
probability density ϱ_ψ, is recovered immediately. But many other measurements reduce to measurements of position. For example, to 'measure the
spin component σ_x' the particle is allowed to pass through a Stern–Gerlach
magnet and we see whether it is deflected up or down, *i.e.* we observe
position at a subsequent time. Thus the quantum statistics of spin
measurements are also reproduced, and so on.

This scheme is readily generalized to many particle systems, within the
framework of nonrelativistic wave mechanics. The wave function is now in
the $3n$-dimensional configuration space

$$\psi(\mathbf{r}_1, \mathbf{r}_2, \ldots, t)$$

and the Schrödinger equation can contain interactions between the
particles. The hidden variables are n-vectors

$$\lambda_1, \lambda_2, \ldots,$$

moving according to

$$(d\lambda_m/dt) = \mathbf{j}_{m\psi}(\lambda_1, \lambda_2, \ldots, t)/\varrho_\psi(\lambda_1, \lambda_2, \ldots, t),$$

$$\varrho_\psi(\lambda_1, \lambda_2, \ldots, t) = |\psi(\lambda_1, \lambda_2, \ldots, t)|^2,$$

$$\mathbf{j}_{m\psi}(\lambda_1, \lambda_2, \ldots, t) = \tfrac{1}{2} \operatorname{Im} \psi^*(\partial/\partial \mathbf{r}_m)\psi |_{r=\lambda}.$$

Again the ensemble corresponding to the quantum-mechanical state has
the λs initially distributed with probability density $|\psi|^2$ in the $3n$-
dimensional space, and this remains so in virtue of the equations of motion.
Thus the quantum statistics of position measurements, and of any
procedure ending up in a position measurement (be it only the observation
of a pointer reading) can be reproduced.

What happens to the hidden variables during and after the measurement is a delicate matter. Note only that a prerequisite for a specification of what happens to the hidden variables would be a specification of what happens to the wave function. But it is just at this point that the notoriously vague 'reduction of the wave packet' intervenes, at some ill-defined time, and we come up against the ambiguities of the usual theory, which for the moment we aim only to reinterpret rather than to replace. It would indeed be very interesting to go beyond this point. But we will not make the attempt here, for we will find a very striking difficulty at the level to which the scheme has been developed already. Before coming to this, a number of instructive features of the scheme are worth indicating.

One such feature is this. We have here a picture in which although the wave has two components, the particle has only position λ. The particle does not 'spin', although the experimental phenomena associated with spin are reproduced. Thus the picture resulting from a hidden-variable account of quantum mechanics need not very much resemble the traditional classical picture that the researcher may, secretly, have been keeping in mind. The electron need not turn out to be a small spinning yellow sphere.

A second way in which the scheme is instructive is in the explicit picture of the very essential role of apparatus. The result of a 'spin measurement', for example, depends in a very complicated way on the initial position λ of the particle and on the strength and geometry of the magnetic field. Thus the result of the measurement does not actually tell us about some property previously possessed by the system, but about something which has come into being in the combination of system and apparatus. Of course, the vital role of the complete physical set-up we learned long ago, especially from Bohr. When it is forgotten, it is more easy to expect that the results of the observations should satisfy some simple algebraic relations and to feel that these relations should be preserved even by the hypothetical dispersion-free states of which quantum-mechanical states may be composed. The model illustrates how the algebraic relations, valid for the statistical ensembles, which are the quantum-mechanical states, may be built up in a rather complicated way. Thus the contemplation of this simple model could have a liberalizing effect on mathematical investigators.

Finally, this simple scheme is also instructive in the following way. Even if the infamous boundary, between classical and quantum worlds, should not go away, but rather become better defined as the theory evolves, it seems to me that some classical variables will remain essential (they may describe 'macroscopic' objects, or they may be finally restricted to apply only to my sense data). Moreover, it seems to me that the present 'quantum theory of measurement' in which the quantum and classical levels interact only fitfully during highly idealized 'measurements' should be replaced by an interaction of a continuous, if variable, character. The eqs. (3) and (4) of the simple scheme form a sort of prototype of a master equation of the world in which classical variables are continuously influenced by a quantum-mechanical state.

4 A difficulty

The difficulty is this. Looking at (4) one sees that the behaviour of a given variable λ_1 is determined not only by the conditions in the immediate neighbourhood (in ordinary three-space) but also by what is happening at all the other positions $\lambda_2, \lambda_3, \ldots$. That is to say, that although the system of equations is 'local' in an obvious sense in the $3n$-dimensional space, it is not at all local in ordinary three-space. As applied to the Einstein–Podolsky–Rosen situation, we find that this scheme provides an explicit causal mechanism by which operations on one of the two measuring devices can influence the response of the distant device. This is quite the reverse of the resolution hoped for by EPR, who envisaged that the first device could serve only to reveal the character of the information already stored in space, and propagating in an undisturbed way towards the other equipment.

The question then arises: can we not find another hidden-variable scheme with the desired local character? It can be shown that this is not possible[7–9]. The demonstration moreover is in no way restricted to the context of nonrelativistic wave mechanics, but depends only on the existence of separated systems highly correlated with respect to quantities such as spin.

Consider again for example the system of two spin-$\frac{1}{2}$ particles. Suppose they have been prepared somehow in such a state that they then move in different directions towards two measuring devices, and that these devices measure spin components along directions $\hat{a}$ and $\hat{b}$ respectively. Suppose that the hypothetical complete description of the initial state is in terms of hidden variables λ with probability distribution $\varrho(\lambda)$ for the given quantum-mechanical state. The result A ($= \pm 1$) of the first measurement can clearly depend on λ and on the setting $\hat{a}$ of the first instrument. Similarly, B can depend on λ and $\hat{b}$. But our notion of locality requires *that A does not depend on $\hat{b}$, nor B on $\hat{a}$*. We then ask if the mean value $P(\hat{a}, \hat{b})$ of the product AB, i.e.

$$P(\hat{a}, \hat{b}) = \int d\lambda \varrho(\lambda) A(\hat{a}, \lambda) B(\hat{b}, \lambda) \tag{5}$$

can equal the quantum-mechanical prediction.

Actually we can, and should, be somewhat more general. The instruments themselves could contain hidden variables[10] which could influence the results. If we average first over these instrument variables, we obtain the representation

$$P(\hat{a}, \hat{b}) = \int d\lambda \varrho(\lambda) \bar{A}(\hat{a}, \lambda) \bar{B}(\hat{b}, \lambda), \tag{6}$$

where the averages $\bar{A}$ and $\bar{B}$ will be independent of $\hat{b}$ and $\hat{a}$, respectively, if *the corresponding distributions of instrument variables are independent of b and a, respectively*, although of course they may depend on $\hat{a}$ and $\hat{b}$, respectively. Instead of

$$A = \pm 1, \quad B = \pm 1, \tag{7}$$

we now have

$$|\bar{A}| \leqslant 1, \quad |\bar{B}| \leqslant 1, \tag{8}$$

and this suffices to derive an interesting restriction on P.

In practice, there will be some occasions on which one or both instruments simply fail to register either way. One might then[11] count A and/or B as zero in defining P, $\bar{A}$, and $\bar{B}$; (8) remains true and the following reasoning remains valid.

Let $\hat{a}'$ and $\hat{b}'$ be alternative settings of the instruments. Then

$$P(\hat{a}, \hat{b}) - P(\hat{a}, \hat{b}') = \int d\lambda \varrho(\lambda) [\bar{A}(\hat{a}, \lambda) \bar{B}(\hat{b}, \lambda) - \bar{A}(\hat{a}, \lambda) \bar{B}(\hat{b}', \lambda)]$$

$$= \int d\lambda \varrho(\lambda) [\bar{A}(\hat{a}, \lambda) \bar{B}(\hat{b}, \lambda)(1 \pm \bar{A}(\hat{a}', \lambda) \bar{B}(\hat{b}', \lambda))]$$

$$- \int d\lambda \varrho(\lambda) [\bar{A}(\hat{a}, \lambda) \bar{B}(\hat{b}', \lambda)(1 \pm \bar{A}(\hat{a}', \lambda) \bar{B}(\hat{b}, \lambda))].$$

Then using (8)

$$|P(\hat{a}, \hat{b}) - P(\hat{a}, \hat{b}')| \leqslant \int d\lambda \varrho(\lambda)(1 \pm \bar{A}(\hat{a}', \lambda) \bar{B}(\hat{b}', \lambda))$$

$$+ \int d\lambda \varrho(\lambda)(1 \pm \bar{A}(\hat{a}', \lambda) \bar{B}(\hat{b}, \lambda)),$$

or

$$|P(\hat{a}, \hat{b}) - P(\hat{a}, \hat{b}')| \leqslant 2 \pm (P(\hat{a}', \hat{b}') + P(\hat{a}', \hat{b})),$$

or more symmetrically

$$|P(\hat{a}, \hat{b}) - P(\hat{a}, \hat{b}')| + |P(\hat{a}', \hat{b}') + P(\hat{a}', \hat{b})| \leqslant 2. \tag{9}$$

With $\hat{a}' = \hat{b}'$ and assuming

$$P(\hat{b}', \hat{b}') = -1, \tag{10}$$

equation (9) yields

$$|P(\hat{a}, \hat{b}) - P(\hat{a}, \hat{b}')| \leqslant 1 + P(\hat{b}', \hat{b}). \tag{11}$$

This is the original form of the result[7]. Note that to realize (10) it is necessary that the equality sign holds in (8), *i.e.* for this case the possibility of the results depending on hidden variables in the instruments can be excluded from the beginning[12].

The more general relation (9) (essentially) was first written by Clauser, Holt, Horne and Shimony[8] for the restricted representation (5).

Suppose now, for example, that the system was in the singlet state of the two spins. Then quantum-mechanically $P(a, b)$ is given by the expectation value in that state

$$\langle \boldsymbol{\sigma}_1 \cdot \hat{a}, \boldsymbol{\sigma}_2 \cdot \hat{b} \rangle = -\hat{a} \cdot \hat{b}. \tag{12}$$

This function has the property (10), but does not at all satisfy (11). With $P(\hat{a}, \hat{b}) = -\hat{a} \cdot \hat{b}$ one finds, for example, that for a small angle between $\hat{b}$ and $\hat{b}'$ the left-hand side of (11) is in general of first order in this angle, while the right-hand side is only of second order. Thus the quantum-mechanical result cannot be reproduced by a hidden-variable theory which is local in the way described.

This result opens up the possibility of bringing the questions that we have been considering into the experimental area. Of course, the situation envisaged above is highly idealized. It is supposed that the system is initially in a known spin state, that the particles are known to proceed towards the instruments, and to be measured there with complete efficiency. The question then is whether the inevitable departures from this ideal situation can be kept sufficiently small in practice that the quantum-mechanical prediction still violates the inequality (9).

In this connection other systems, for example the two-photon system[8] or the two-kaon system[13], may be more promising than that of two-spin $\frac{1}{2}$ particles. A very serious study of the photon case will be reported to this meeting by Shimony. The experiment described by him, and now under way, is not sufficiently close to the ideal to be conclusive for a quite determined advocate of hidden variables. However, for most a confirmation of the quantum-mechanical predictions, which is only to be expected given the general success of quantum mechanics[14], would be a severe discouragement.

Notes and references

1 A. Einstein, B. Podolsky and N. Rosen: *Phys. Rev.*, **47**, 777 (1935).

2 D. Bohm: *Quantum Theory* (Englewood Cliffe, N.J., 1951).

3 J. von Neumann: *Mathematische Grundlagen der Quantenmechanik*, Berlin (1932) (English translation (Princeton, 1955)).

4 For an analysis of some of these schemes, see: J. S. Bell: *Rev. Mod. Phys.*, **38**, 447 (1966). This considers in particular the result of J. M. Jauch and C. Piron, *Helv. Phys. Acta*, **36**, 827 (1963) and the stronger form of von Neumann's result consequent, as observed by Jauch, on the work of A. M. Gleason, *J. Math. and Mech.*, **6**, 885 (1957). This corollary of Gleason's work was subsequently set out by S. Kochen and E. P. Specker, *J. Math. and Mech.*, **17**, 59 (1967). Other impossibility proofs have been given by S. P. Gudder, *Rev. Mod. Phys.*, **40**, 229 (1968) and by B. Misra, *Nuovo Cimento*, **47**, 843 (1967); both of these authors remark on the limited nature of their results. On the question of impossibility proofs, see also D. Bohm and J. Bub, *Rev. Mod. Phys.*, **38**, 453 (1966); **40**, 232 (1968); J. M. Jauch and C. Piron, *Rev. Mod. Phys.*, **40**, 228 (1968) and J. E. Turner, *J. Math. Phys.*, **9**, 1411 (1968).

5 L. de Broglie gives a documented account of the early development in L. de Broglie: *Physicien et Penseur*, p. 465, Paris (1953).

6 D. Bohm: *Phys. Rev.*, **85**, 166, 180 (1952). For hidden variable schemes see also the review of H. Friestadt. *Suppl. Nuovo Cimento*, **5**, 1 (1967) and later work by D. Bohm and J. Bub, *Rev. Mod. Phys.*, **38**, 470 (1966) and S. P. Gudder *J. Math. Phys.*, **11**, 431 (1970).

7 J. S. Bell: *Physics*, **1**, 195 (1964).

8 J. F. Clauser, M. A. Horne, A. Shimony and R. A. Holt: *Phys. Rev. Lett.*, **26**, 880 (1969).

9 E. P. Wigner: *Am. J. Phys.*, **38**, 1005 (1970).

10 We speak here as if the instruments responded in a deterministic way when all variables, hidden or nonhidden, are given. Clearly (6) is appropriate also for *indeterminism* with a certain local character.

11 This is a suggestion of J. A. Crawford.

12 This was the procedure, as regards the ideal case (12), in ref.[7]. However in that reference the subsequent discussion of the nonideal case started again from the restricted representation (5). This was quite arbitrary. But the reasoning of that section, used again here, goes through with the more general (6). In this connection, I am indebted to J. A. Crawford for a stimulating correspondence.

13 T. B. Day: *Phys. Rev.*, **121**, 1204 (1961); D. R. Inglis: *Rev. Mod. Phys.*, **33**, 1, (1961). Note that the spontaneous decay times of the two kaons, because they cannot be set at the will of the experimenter, are not to be regarded as analogous to the setting *a* and *b* of the Stern–Gerlach magnets. The thicknesses of a pair of slabs of matter placed in the lines of flight would be more relevant. I am told by Prof. B. d'Espagnat that the rapid decay of the short-lived kaon is a major obstacle to devising a critical experiment.

14 The helium atom, essentially a pair of spin-$\frac{1}{2}$ particles, is a system for which quantum mechanics is strikingly successful. See, for example, H. A. Bethe and E. E. Salpeter: *Handbuch der Physik*, Vol. 35 p. 88, Berlin (1957).

5

Subject and object

The subject–object distinction is indeed at the very root of the unease that many people still feel in connection with quantum mechanics. *Some* such distinction is dictated by the postulates of the theory, but exactly *where* or *when* to make it is not prescribed. Thus in the classic treatise[1] of Dirac we learn the fundamental propositions:

> ... any result of a measurement of a real dynamical variable is one of its eigenvalues ...,
> ... if the measurement of the observable ξ for the system in the state corresponding to $|x\rangle$ is made a large number of times, the average of all the results obtained will be $\langle x|\xi|x\rangle$ ...,
> ... a measurement always causes the system to jump into an eigenstate of the dynamical variable that is being measured ....

So the theory is fundamentally about the results of 'measurements', and therefore presupposes in addition to the 'system' (or object) a 'measurer' (or subject). Now must this subject include a person? Or was there already some such subject–object distinction before the appearance of life in the universe? Were some of the natural processes then occurring, or occurring now in distant places, to be identified as 'measurements' and subjected to jumps rather than to the Schrödinger equation? Is 'measurement' something that occurs all at once? Are the jumps instantaneous? And so on.

The pioneers of quantum mechanics were not unaware of these questions, but quite rightly did not wait for agreed answers before developing the theory. They were entirely justified by results. The vagueness of the postulates in no way interferes with the miraculous accuracy of the calculations. Whenever necessary a little more of the world can be incorporated into the object. In extremis the subject–object division can be put somewhere at the 'macroscopic' level, where the practical adequacy of classical notions makes the precise location quantitatively unimportant. But although quantum mechanics can account for these classical features of the macroscopic world as very (very) good approximations, it cannot do more than that.[2] The snake cannot completely swallow itself by the tail. This awkward fact remains: the theory is only *approximately* unambiguous, only *approximately* self-consistent.

It would be foolish to expect that the next basic development in theoretical physics will yield an accurate and final theory. But it is interesting to speculate on the possibility that a future theory will not be *intrinsically* ambiguous and approximate. Such a theory could not be fundamentally about 'measurements', for that would again imply incompleteness of the system and unanalyzed interventions from outside. Rather it should again become possible to say of a system not that such and such may be *observed* to be so but that such and such *be* so. The theory would not be about '*observa*bles' but about '*bea*bles'. These beables need not of course resemble those of, say, classical electron theory; but at least they should, on the macroscopic level, yield an image of the everyday classical world[4], for 'it is decisive to recognize that, however far the phenomena transcend the scope of classical physical explanation, the account of all evidence must be expressed in classical terms'.[5]

By 'classical terms' here Bohr is not of course invoking particular nineteenth-century theories, but refers simply to the familiar language of everyday affairs, including laboratory procedures, in which objective properties – *beables* – are assigned to objects. The idea that quantum mechanics is primarily about 'observables' is only tenable when such beables are taken for granted. Observables are *made* out of beables. We raise the question as to whether the beables can be incorporated into the theory with more precision than has been customary.

Many people must have thought along the following lines. Could one not just promote *some* of the 'observables' of the present quantum theory to the status of beables? The beables would then be represented by linear operators in the state space.[6] The values which they are allowed to *be* would be the eigenvalues of those operators. For the general state the probability of a beable *being* a particular value would be calculated just as was formerly calculated the probability of *observing* that value. The proposition about the jump of state consequent on measurement could be replaced by: when a particular value is attributed to a beable, the state of the system reduces to a corresponding eigenstate. It is the main object of this note to set down some remarks on this programme. Perhaps it is only because they are quite trivial that I have not seen them set down already.

The state vector (or density matrix) in what follows will always be that of the Heisenberg picture: all time dependence is in the operators and the state refers not to a single time but to a whole history. This permits us, if we wish, to define the 'system' under study simply as a limited space-time region. This seems a less intrinsically ambiguous and unrealistic way than any other I can think of to separate off a part of the world from the rest. Of course, one could try to think of the world as a whole, but it is less intimidating to think of only a part. In the approach[8] known as the 'theory of local observables' a Heisenberg state (pure or mixed) can indeed be attributed to any limited region of space-time. It gives, roughly speaking, the expectation value of all functions of the Heisenberg field operators with space-time arguments in that region. If something like a Lorentz-invariant causal connection between field operators is postulated then the region of relevance of the state vector can be extended by including all points whose forward or backward light cones pass entirely through the original region, as in Fig. 1. It is then the Heisenberg state of the extended region which reduces, whenever a 'local beable' in that region is attributed a particular value, to its projection in the subspace with the given eigenvalue. Whatever the particular space-time location of the beable considered, there is no question of any particular space-time location of the associated state reduction, which is coextensive with the whole history of the system under study.

Whereas 'measurement' was a dynamical intervention, from somewhere outside, with dynamical consequences, it is clear that 'attribution' must be regarded as a purely conceptual intervention. It is made, say, by a theorist rather than an experimenter; he is quite remote in space and time from the action, and simply shifts his attention from the whole of a statistical ensemble to a sub-ensemble. It follows that attributing a particular value to some beable cannot change particular values already attributed to some

Fig. 1.

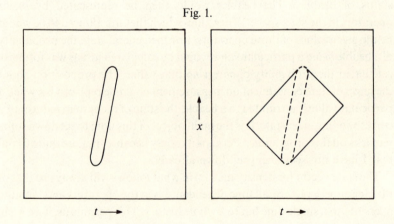

other beables. It follows that only those states can be allowed which are simultaneously eigenstates of all beables, or superpositions of such states. Moreover, we need only consider incoherent superpositions, for the beables, unable to induce transitions between different eigenstates, are insensitive to any coherence. Now the beables may not be a complete set, and a list of their eigenvalues may not characterize a state completely. However, the converse is true: when a particular member state of the incoherent superposition is specified, definite values are specified for all beables. Thus the theory is of deterministic hidden-variable type, with the Heisenberg state playing the role of hidden variable. When this state, which may originally refer only to the limited region in the figure, is specified, all beables in the extended region are determined.

I suspect that a stronger conclusion would be possible, that one cannot in fact find interesting candidates for beables in interesting quantum mechanical systems. But my own indications in this direction seem to me unnecessarily elaborate and I will not attempt to present them here. The preliminary conclusion is in a way more striking. In the basic propositions quoted from Dirac there was in fact another element, in addition to the vague subjectivity, which could have disturbed a nineteenth-century theorist. That is the *statistical undeterministic* character of the basic notions. In following what seemed to be a minimal programme for restoring objectivity, we were obliged to restore determinism also.

Notes and references

1 P. A. M. Dirac, *The Principles of Quantum Mechanics*.
2 In this connection there are many very relevant investigations involving considerations which may be roughly identified by the words 'ergodicity' or 'irreversibility'. They tend to show that the effect of wave packet reduction associated with macroscopic observation is macroscopically negligible. (Or it may even be shown that the effect is accurately zero in some hypothetical limit: e.g., K. Hepp[3] takes infinite time.) The relevance of these investigations is of course to the question of the sufficient unambiguity of the theory for practical purposes, and not at all to the question of principle considered here.
3 K. Hepp, *Helv. Phys. Acta* **45**, 237 (1972).
4 A more extreme position would be that the beables need refer only to mental events.
5 N. Bohr.
6 Such beables would be related to the 'classical observables' of Jauch and Piron (see for example the contributions of these authors in *Foundations of Quantum Mechanics*, Proceedings of the International School of Physics 'Enrico Fermi', Course IL, Academic Press, New York, 1971; also H. Primas[7]). However, these authors (*loc. cit.* and private communications) intended their 'classical observables' to refer only to 'apparatus' while not in interaction with 'quantum systems' and perhaps to be only approximately 'classical'. Here we wish to avoid any arbitrary

division of the world into 'systems' and 'apparatus', and any arbitrary limitation on the range and duration of interactions, and are concerned with the question of principle and not with that of practical approximation.

7 H. Primas, *Advanced Quantum Chemistry of Large Molecules*, Vol. 1: 'Concepts and Kinematics of Quantum Mechanics of Large Molecular Systems', Academic Press, New York (1973), and preprint (July 1972).

8 See, for example: R. Haag, in *Lectures on Elementary Particles and Quantum Field Theory*, 1970 Brandeis Lectures, Editors S. Deser, M. Grisaru and H. Pendleton, M.I.T. Press (1970). In this theory the over-all system need not be finite. The idea that the measurement problem might be significantly different in such a context has sometimes been expressed.[3,7,9]

9 See, for example, the preface to B. d'Espagnat's *Conceptual Foundations of Quantum Mechanics*, Benjamin, New York (1971).

6

On wave packet reduction in the Coleman–Hepp model

1 Introduction

In a very elegant and rigorous paper[1], K. Hepp has discussed quantum measurement theory. He uses the C^* algebra description of infinite quantum systems. Here an attempt is made to give a more popular account of some of his reasoning. Such an attempt seems worthwhile because many people not familiar with the C^* algebra approach, and even somewhat intimidated by it, have been intrigued by the following statement in Hepp's abstract:

> In several explicitly soluble models, the measurement leads to macroscopically different 'pointer positions' and to a rigorous 'reduction of the wave packet' with respect to all local observables.

This could look like a clean solution at last to the infamous measurement problem[2]. But it is not so, nor thought by Hepp to be so. Here we will take one[4] of his models and analyse it in elementary text-book terms. It will be insisted that the 'rigorous reduction' does not occur in physical time but only in an unattainable mathematical limit. It will be argued that the distinction is an important one.

We will work at first in the Schrödinger picture, but later, with the extension to relativistic systems in mind, it will be argued that such considerations become particularly clear in the Heisenberg picture.

2 Model

The model is the following. The 'apparatus' is a semi-infinite linear array of spin-$\frac{1}{2}$ particles, fixed at positions $x = 1, 2, \ldots$. The 'system' is a moving spin-$\frac{1}{2}$ particle, with position co-ordinate x and spin operators $\boldsymbol{\sigma}_0 (\equiv \sigma_0^1, \sigma_0^2, \sigma_0^3)$; it is the third component σ_0^3 which is to be 'measured'. The combined system is described by a wave function, where all σ_n take values ± 1,

$$\psi(t, x, \sigma_0, \sigma_1, \sigma_2, \ldots)$$

in a representation where all σ_n^3 are diagonal:

$$\sigma_n^3 \psi(t, x, \sigma_0, \sigma_1, \sigma_2, \ldots) = \sigma_n \psi(t, x, \sigma_0, \sigma_1, \sigma_2, \ldots). \tag{1}$$

The Hamiltonian is taken to be

$$H = \frac{1}{i}\frac{\partial}{\partial x} + \sum_{n=1}^{\infty} V(x - n)\sigma_n^1(\tfrac{1}{2} - \tfrac{1}{2}\sigma_0^3). \tag{2}$$

Note that the 'kinetic energy' here is linear rather than quadratic in the particle momentum $p = (1/i)(\partial/\partial x)$. This has the convenience that free particle wave packets do not diffuse; they just move without change of form, and with unit velocity, in the positive x-direction. The interaction V is supposed to have 'compact support' – i.e., to be zero beyond some range r:

$$V(x) = 0 \quad \text{for } |x| > r. \tag{3}$$

It is also supposed, for reasons that will appear, that

$$\int_{-\infty}^{\infty} \mathrm{d}x\, V(x) = \frac{\pi}{2}. \tag{4}$$

The Schrödinger equation

$$\frac{\partial \psi}{\partial t} = -iH\psi$$

is readily solved

$$\psi(t, x, \sigma_0, \ldots) = \prod_{n=1}^{\infty} \exp\left[-iF(x - n)\sigma_n^1(\tfrac{1}{2} - \tfrac{1}{2}\sigma_0^3)\right]\phi(x - t, \sigma_0, \ldots) \tag{5}$$

where ϕ is arbitrary and

$$F(x) = \int_{-\infty}^{x} \mathrm{d}y\, V(y). \tag{6}$$

Note that

$$\left.\begin{array}{ll} F(x) = 0 & \text{for } x < -r \\ F(x) = \pi/2 & \text{for } x > +r. \end{array}\right\} \tag{7}$$

Consider in particular states in which initially the lattice spins are all up and the moving spin is either up or down:

$$\left.\begin{array}{l} \psi_+(t, x, \ldots) = \chi(x - t)\psi_+(\sigma_0)\displaystyle\prod_{n=1}^{\infty}\psi_+(\sigma_n) \\[2mm] \psi_-(t, x, \ldots) = \chi(x - t)\psi_-(\sigma_0)\displaystyle\prod_{n=1}^{\infty}\psi_+'(\sigma_n, x - n) \end{array}\right\} \tag{8}$$

where

$$\left.\begin{array}{l} \psi_\pm(\sigma) = \delta_{\sigma\mp1} \\ \psi'_+(\sigma_n, x-n) = \exp\left[-iF(x-n)\sigma_n^1\right]\psi_+(\sigma_n). \end{array}\right\} \tag{9}$$

Note that in virtue of (7)

$$\left.\begin{array}{ll} \psi'_+(\sigma_n, x-n) = \psi_+(\sigma_n) & \text{for } x-n < -r \\ \qquad\qquad\quad = -i\psi_-(\sigma_n) & \text{for } x-n > +r. \end{array}\right\} \tag{10}$$

Let us suppose that the wave packet χ has compact support:

$$\chi(x) = 0 \quad \text{for } |x| > w. \tag{11}$$

Then, from (10) we can use in (8)

$$\left.\begin{array}{ll} \psi'_+(\sigma_n, x-n) = \psi_+(\sigma_n) & \text{for } n > t+r+w \\ \psi'_+(\sigma_n, x-n) = -i\psi_-(\sigma_n) & \text{for } n < t-r-w. \end{array}\right\} \tag{12}$$

Thus (8) has the interpretation that when the system spin is up nothing happens to the apparatus spins, but when the system spin is down each apparatus spin in turn is flipped from up to down.

Hepp's 'macroscopic pointer position' can be defined here by considering the limit $M \to \infty$ of

$$C_M = \frac{1}{M} \sum_{n=1}^M \sigma_n^3. \tag{13}$$

Clearly

$$\operatorname*{Lim}_{M\to\infty}\left(\operatorname*{Lim}_{t\to\infty}(\psi_\pm, C_M\psi_\pm)\right) = \pm 1. \tag{14}$$

So we have his 'macroscopically different pointer positions'. From the fact that the two states have different values here (for what Hepp calls a 'classical observable', involving infinitely many of the basic operators σ) Hepp infers that

$$\operatorname*{Lim}_{t\to\infty}(\psi_\pm, Q\psi_\mp) = 0 \tag{15}$$

for any 'local observable' Q – i.e., one constructed from a *finite* number of σs. This is plausible in general because such a difference means, loosely speaking, that the two states differ significantly at infinitely many lattice points, and so remain mutually orthogonal after any operation involving only finitely many lattice points. In this particular case, we see explicitly from (12) that if a particular Q involves only $(\sigma_0, \sigma_1 \cdots \sigma_N)$ then

$$(\psi_\pm, Q\psi_\mp) = 0 \quad \text{for } t > 1 + N + r + w \tag{16}$$

which includes (15).

The result (15) is the 'rigorous reduction of the wave packet'. If the 'local observables' Q (as distinct in particular from the 'classical observables') are thought of as those which can in principle actually be observed, then the vanishing of their matrix elements between the two states means that coherent superpositions of ψ_+ and ψ_- cannot be distinguished from incoherent mixtures thereof. In quantum measurement theory such elimination of coherence is the philosopher's stone. For with an incoherent mixture specialization to one of its components can be regarded as a purely mental act, the innocent selection of a particular subensemble, from some total statistical ensemble, for particular further study.

We insist, however, that $t = \infty$ never comes, so that the wave packet reduction never happens. The mathematical limit $t \to \infty$ is of physical relevance only in so far as it suggests what might be true, or nearly so, for large t. The result (15) (and more sharply, in this particular case, (16)) shows that any *fixed* observable Q will eventually give a very poor (zero, in this case) measure of the persisting coherence. But nothing forbids the use of different observables as time goes on. Consider for example the unitary operator

$$z = \sigma_0^1 \prod_{n=1}^{N(t-r-w)} \sigma_n^2 \tag{17}$$

where $N(t)$ is the largest integer smaller than t. The increasing string of factors here serves to unflip the flipped spins, so that

$$(\psi_+, z\psi_-) = \int dx \, |\chi(x-t)|^2 \prod_{N(t-r-w)}^{N(t+r+w)} (\psi_+(\sigma_n), \psi'_+(\sigma_n, x-n)) \tag{18}$$

becomes a periodic function of t. Trivially,

$$(\psi_+, z\psi_+) = (\psi_-, z\psi_-) = 0. \tag{19}$$

Thus in the Hermitean operators z we have a sequence of local observables whose matrix elements

$$(\psi_\mp, z\psi_\pm) \tag{20}$$

do *not* approach zero. So long as nothing, in principle, forbids consideration of such arbitrarily complicated observables, it is not permitted to speak of wave packet reduction. While for any given observable one can find a time for which the unwanted interference is as small as you like, for any given time one can find an observable for which it is as big as you do *not* like.

3 Heisenberg picture

Consider now the Heisenberg picture[5], in which the states are time-independent and the operators vary. The Heisenberg equations of motion are in general

$$\dot{Q}(t) = [Q(t), -iH]$$

and in particular

$$\dot{x}(t) = 1$$

$$\dot{\boldsymbol{\sigma}}_0(t) = -\left(\sum_{n=1}^{\infty} V(x(t)-n)\sigma_n^1(t)\right)\hat{\mathbf{k}} \times \boldsymbol{\sigma}_0(t)$$

$$\dot{\boldsymbol{\sigma}}_n(t) = +\left(\sum_{n=1}^{\infty} V(x(t)-n)\right)(1-\sigma_0^3(t))\hat{\mathbf{i}} \times \boldsymbol{\sigma}_n(t)$$

where $\hat{\mathbf{i}}$ and $\hat{\mathbf{k}}$ are unit vectors in the 1 and 3 directions. Now we could solve these equations forward in time to find subsequent values in terms of initial values, and then to say again what has been said above. But we wish to note rather that the equations can be solved *backwards* in time, to express operators at some initial time in terms of those at any later time. For example, we find

$$\sigma_0^1(0) = \sigma_0^1(t)\cos\theta(t) - \sigma_0^2(t)\sin\theta(t) \tag{21}$$

where

$$\theta(t) = \sum_{n=1}^{\infty} \{F(x(t)-n) - F(x(t)-t-n)\}\sigma_n^1(t). \tag{22}$$

Between states which satisfy the Schrödinger equation, matrix elements of σ_0^1 at time zero are equal to the corresponding matrix elements at time t of the combination of observables on the right-hand side of (21). Thus this combination serves the same purpose as that of (17), of giving a constant measure to the persisting coherence – in this case whatever coherence could initially be measured by σ_0^1. It is not, of course, the same construction as (17), and in fact it explicitly invokes $x(t)$, as well as $\sigma_n(t)$, as an observable. But why not?

We note in passing that in the Heisenberg picture there is no complication in considering mixed rather than pure states. Whatever coherence shows up at time 0 in the expectation value of an operator $Q(0)$, will persist and show up at later times in the expectation value of the corresponding combination of $Q(t)$. In this picture the persistence of coherence is directly related to the deterministic character of the Heisenberg equations of motion. This operates backwards as well as forwards in time, and requires a given $Q(0)$ to be some combination of the set $Q(t)$ with any given t.

As written, the summation in (22) is infinite. But for any given wave packet $\chi(x)$, of compact support, it can be terminated without error at some sufficiently large n, growing with time. This is because of (7), which requires F to vanish for large negative arguments. Thus, loosely speaking, the evidence for coherence remains at any finite time in a finite region of the lattice. This will not be generally true in nonrelativistic models. It is associated with the use of interactions and wave packets of compact support, and with the existence in the model of a limiting – indeed universal – velocity, which was taken to be unity.

In *relativistic* theories, however, we again have a limiting velocity, that of light – at least if we have flat unquantized space-time and can avoid the pathologies of Velo and Zwanziger[7]. The local observables in an initial space-time region are then presumably determined by those contained subsequently in a region obtained from the original by expanding its space boundaries with the velocity of light. Presumably the exact formulation of this notion is to be found in the 'primitive causality' of Haag[8]. In so far as it applies we see again that any coherence associated with the initial region must persist, and be detectable subsequently in a bigger but finite region by using the appropriate combination of observables in that region.

4 Conclusion

Clearly there is no room for disagreement about simple mathematics. But there may be disagreement about the physical significance of it. Hepp clearly considers the limit $t \to \infty$ very relevant, while he does 'not, however, accept the ergodic mean as a fundamental solution to the problem of the reduction of wave packets'. In my opinion neither of these approaches provides a *fundamental* solution, but both are quite valuable for indicating how the difference between reducing the wave packet at one time rather than another is extremely hard to see *in practice*. Moreover, both indicate this on the same ground – that the observation of arbitrarily complicated observables, while not excluded in principle, is not possible in practice. It remains true that, whenever it is done, the wave packet reduction is not compatible with the linear Schrödinger equation. And yet at some not-well-specified time, such a reduction is supposed to occur[9]: '⋯a measurement always causes the system to jump into an eigenstate of the dynamical variable that is measured ···'.

The continuing dispute about quantum measurement theory is not between people who disagree on the results of simple mathematical manipulations. Nor is it between people with different ideas about the actual practicality of measuring arbitrarily complicated observables. It is between people who view with different degrees of concern or complacency the following fact: so long as the wave packet reduction is an essential component, and so long as we do not know exactly when and how it takes over from the Schrödinger equation, we do not have an exact and unambiguous formulation of our most fundamental physical theory.

Acknowledgements

I thank B. d'Espagnat, V. Glaser, K. Hepp and H. Ruegg for useful discussions.

Notes and references

1 K. Hepp, *Helv. Phys. Acta* **45**, 237 (1972).
2 For a general survey, see, for example, d'Espagnat[3].
3 B. d'Espagnat, *Conceptual Foundations of Quantum Mechanics*, Benjamin, Addison-Wesley, Reading, Mass. (1971).
4 Note that Hepp considers several other models, making points not presented here, in particular concerning the possibility of 'catastrophic' time evolutions.
5 The use of the Heisenberg picture in quantum measurement theory has been advocated, for different reasons, by B. S. De Witt[6].
6 B. S. De Witt, in *Foundations of Quantum Mechanics, Proceedings of International School of Physics Enrico Fermi*, Course 49, edited by B. d'Espagnat. Academic Press, N.Y. (1971).
7 G. Velo and D. Zwanziger, *Phys. Rev.* **188**, 2218 (1969).
8 R. Haag, in *Lectures on Elementary Particles and Quantum Field Theory*, 1970 Brandeis Lectures, edited by S. Deser, M. Grisaru and H. Pendleton. M.I.T. Press, (1970).
9 P. A. M. Dirac, *Quantum Mechanics*.

7

The theory of local beables

Introduction: the theory of local beables

This is a pretentious name for a theory which hardly exists otherwise, but which ought to exist. The name is deliberately modelled on 'the algebra of local observables'. The terminology, *be*-able as against *observ*-able, is not designed to frighten with metaphysic those dedicated to realphysic. It is chosen rather to help in making explicit some notions already implicit in, and basic to, ordinary quantum theory. For, in the words of Bohr[1], 'it is decisive to recognize that, however far the phenomena transcend the scope of classical physical explanation, the account of all evidence must be expressed in classical terms'. It is the ambition of the theory of local beables to bring these 'classical terms' into the equations, and not relegate them entirely to the surrounding talk.

The concept of 'observable' lends itself to very precise *mathematics* when identified with 'self-adjoint operator'. But physically, it is a rather woolly concept. It is not easy to identify precisely which physical processes are to be given the status of 'observations' and which are to be relegated to the limbo between one observation and another. So it could be hoped that some increase in precision might be possible by concentration on the *beables*, which can be described in 'classical terms', because they are there. The beables must include the settings of switches and knobs on experimental equipment, the currents in coils, and the readings of instruments. 'Observables' must be *made*, somehow, out of beables. The theory of local beables should contain, and give precise physical meaning to, the algebra of local observables.

The word 'beable' will also be used here to carry another distinction, that familiar already in classical theory between 'physical' and 'non-physical' quantities. In Maxwell's electromagnetic theory, for example, the fields $\mathbf{E}$ and $\mathbf{H}$ are 'physical' (beables, we will say) but the potentials $\mathbf{A}$ and ϕ are 'non-physical'. Because of gauge invariance the same physical situation can be described by very different potentials. It does not matter that in

Coulomb gauge the scalar potential propagates with infinite velocity. It is not really supposed to *be* there. It is just a mathematical convenience.

One of the apparent non-localities of quantum mechanics is the instantaneous, over all space, 'collapse of the wave function' on 'measurement'. But this does not bother us if we do not grant beable status to the wave function. We can regard it simply as a convenient but inessential mathematical device for formulating correlations between experimental procedures and experimental results, i.e., between one set of beables and another. Then its odd behaviour is as acceptable as the funny behaviour of the scalar potential of Maxwell's theory in Coulomb gauge.

We will be particularly concerned with *local* beables, those which (unlike for example the total energy) can be assigned to some bounded space-time region. For example, in Maxwell's theory the beables local to a given region are just the fields **E** and **H**, in that region, and all functionals thereof. It is in terms of local beables that we can hope to formulate some notion of local causality. Of course we may be obliged to develop theories in which there *are* no strictly local beables. That possibility will not be considered here.

1 Local determinism

In Maxwell's theory, the fields in any space-time region 1 are determined by those in any space region V, at some time t, which fully closes the backward light cone of 1 (Fig. 1). Because the region V is limited, localized, we will say the theory exhibits *local determinism*. We would like to form some notation of *local causality* in theories which are not deterministic, in which the correlations prescribed by the theory, for the beables, are weaker.

Fig. 1.

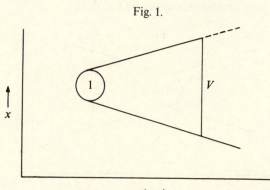

2 Local causality

Consider a theory in which the assignment of values to some beables Λ implies, not necessarily a particular value, but a probability distribution, for another beable A. Let

$$\{A|\Lambda\}$$

denote the probability of a particular value A given particular values Λ. Let A be localized in a space-time region 1. Let B be a second beable localized in a second region 2 separated from 1 in a spacelike way (Fig. 2). Now my intuitive notion of local causality is that events in 2 should not be 'causes' of events in 1, and vice versa. But this does not mean that the two sets of events should be uncorrelated, for they could have common causes in the overlap of their backward light cones. It is perfectly intelligible then that if Λ in (1) does not contain a complete record of events in that overlap, it can be usefully supplemented by information from region 2. So in general it is expected that

$$\{A|\Lambda, B\} \neq \{A|\Lambda\} \tag{1}$$

However, in the particular case that Λ contains already a *complete* specification of beables in the overlap of the two light cones, supplementary information from region 2 could reasonably be expected to be redundant. So, with some change of notation, we formulate local causality as follows.

Let N denote a specification of *all* the beables, of some theory, belonging to the overlap of the backward light cones of spacelike separated regions 1 and 2. Let Λ be a specification of some beables from the remainder of the backward light cone of 1, and B of some beables in the region 2. Then in a *locally causal theory*

$$\{A|\Lambda, N, B\} = \{A|\Lambda, N\} \tag{2}$$

whenever both probabilities are given by the theory.

Fig. 2.

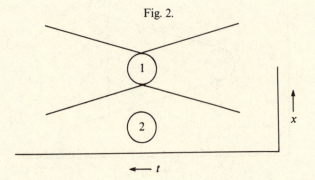

3 Quantum mechanics is not locally causal

Ordinary quantum mechanics, even the relativistic quantum field theory, is not locally causal in the sense of (2). Suppose, for example, we have a radioactive nucleus which can emit a single α-particle, surrounded at a considerable distance by α-particle counters. So long as it is not specified that some *other* counter registers, there is a chance for a particular counter that *it* registers. But if it is specified that some other counter does register, even in a region of space-time outside the relevant backward light cone, the chance that the given counter registers is zero. We simply do not have (2). Could it be that here we have an incomplete specification of the beables, N? Not so long as we stick to the list of beables recognized in ordinary quantum mechanics – the settings of switches and knobs and currents needed to prepare the initial unstable nucleus. For these are completely summarized, in so far as they are relevant for predictions about counter registering, in so far as such predictions are possible in quantum mechanics, by the wave function.

But could it not be that quantum mechanics is a fragment of a more complete theory, in which there are other ways of using the given beables, or in which there are additional beables – hitherto 'hidden' beables? And could it not be that this more complete theory has local causality? Quantum mechanical predictions would then apply not to given values of all the beables, but to some probability distribution over them, in which the beables recognized as relevant by quantum mechanics are held fixed. We will investigate this question, and answer it in the negative.

4 Locality inequality[2-25]

Consider a pair of beables A and B, belonging respectively to regions 1 and 2 with spacelike separation, which happen by definition to have the

Fig. 3.

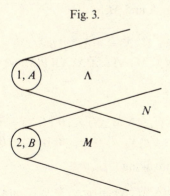

property

$$|A| \leqslant 1 \quad |B| \leqslant 1 \tag{3}$$

Consider the situation in which beables Λ, M, N are specified, where N is a *complete* specification of the beables in the overlap of the light cones, and Λ and M belong respectively to the remainders of the two light cones (Fig. 3).

Consider the joint probability distribution

$$\{A, B | \Lambda, M, N\} \tag{4}$$

By a standard rule of probability, it is equal to

$$\{A | \Lambda, M, N, B\} \{B | \Lambda, M, N\} \tag{5}$$

which, by (2), is the same as

$$\{A | \Lambda, N\} \{B | M, N\} \tag{6}$$

This says simply that correlations between A and B can arise only because of common causes N.

Consider now the expectation value of the product AB

$$p(\Lambda, M, N) = \sum_{A,B} AB \{A | \Lambda, N\} \{B | M, N\} \tag{7}$$

(where the summation stands also, if necessary, for integration)

$$= \bar{A}(\Lambda, N) \bar{B}(M, N) \tag{8}$$

where $\bar{A}$ and $\bar{B}$ are functions of the variables indicated, and

$$|\bar{A}| \leqslant 1 \quad |\bar{B}| \leqslant 1 \tag{9}$$

for all values of the arguments. Let Λ' and M' be alternative specifications, of the same regions, to Λ and M.

$$\left. \begin{array}{l} p(\Lambda, M, N) \pm p(\Lambda, M', N) = A(\Lambda, N)[\bar{B}(M, N) \pm \bar{B}(M', N)] \\ p(\Lambda', M, N) \pm p(\Lambda', M', N) = \bar{A}(\Lambda', N)[\bar{B}(M, N) \pm \bar{B}(M', N)] \end{array} \right\} \tag{10}$$

whence, using (9),

$$\left. \begin{array}{l} |p(\Lambda, M, N) \pm p(\Lambda, M', N)| \leqslant |\bar{B}(M, N) \pm \bar{B}(M', N)| \\ |p(\Lambda', M, N) \pm p(\Lambda', M', N)| \leqslant |\bar{B}(M, N) \pm \bar{B}(M', N)| \end{array} \right\} \tag{11}$$

so that finally, again invoking (9),

$$|p(\Lambda, M, N) \pm p(\Lambda, M', N)| + |p(\Lambda', M, N) \mp p(\Lambda', M', N)| \leqslant 2 \tag{12}$$

Suppose now the specifications Λ, M, N are each given in two parts

$$\Lambda \equiv (a, \lambda)$$
$$M \equiv (b, \mu)$$
$$N \equiv (c, v)$$

where we are particularly interested in the dependence on a, b, c, while λ, μ, v, are averaged over some probability distributions – which may depend on a, b, c. In the comparison with quantum mechanics, we will think of a, b, c, as variables which specify the experimental set-up in the sense of quantum mechanics, while λ, μ, v, are in that sense either hidden or irrelevant. Define

$$P(a, b, c) = \overline{p((a, \lambda), (b, \mu), (c, v))} \qquad (13)$$

where the bar denotes the averaging over (λ, μ, v) just described. Now applying again the locality hypothesis (3), the distribution of λ and v must be independent of b, μ – the latter being outside the relevant backward light cones. So

$$|P(a, b, c) \pm P(a, b', c)| \leqslant \overline{|p((a, \lambda), (b, \mu), (c, v)) \pm p((a, \lambda), (b', \mu'), (c, v))|} \quad (14)$$

– because the mod of the average is less than the average of the mod. In the same way

$$|P(a', b, c) \mp P(a', b', c)| \leqslant \overline{|p((a', \lambda'), (b, \mu), (c, v)) \mp p((a', \lambda'), (b', \mu'), (c, v))|}$$

$$(15)$$

Finally then, from (14), (15) and (12),

$$|P(a, b, c) \mp P(a, b', c)| + |P(a', b, c) \pm P(a', b', c)| \leqslant 2 \qquad (16)$$

5 Quantum mechanics

Quantum mechanics, however, gives certain correlations which *do not satisfy* the locality inequality (16).

Suppose, for example, a neutral pion is produced, by some experimental device, in some small space-time region 3. It quickly decays into a pair of photons. Suppose we have photon counters in space-time regions 1 and 2 so located with respect to 3 that when one photon falls on 1, the second falls (or nearly always does) on 2. If the π^0 is at rest the counters must be equally far away in opposite directions and their sensitive times appropriately delayed. Of course, both photons will often miss both counters. Suppose finally that both counters are behind filters which pass only photons with specified linear polarization, say at angles θ and ϕ respectively to some plane containing the axis joining the two counters.

Let us calculate according to quantum mechanics the probability of the various possible responses of the counters. If $|\theta\rangle$ denotes a photon linearly polarized at an angle θ, then for the photons going towards the counters the combined spin state is

$$|s\rangle = \frac{1}{\sqrt{2}}|0\rangle\left|\frac{\pi}{2}\right\rangle - \frac{1}{\sqrt{2}}\left|\frac{\pi}{2}\right\rangle|0\rangle \qquad (17)$$

where first and second kets in each term refer to the photons going towards regions 1 and 2, respectively. This form is dictated by considerations of parity and angular momentum. The probability that such photons pass the filters is then proportional to

$$\frac{1}{2}|\langle\theta|0\rangle\left\langle\phi\left|\frac{\pi}{2}\right\rangle - \left\langle\theta\left|\frac{\pi}{2}\right\rangle\langle\phi|0\rangle\right|^2\right.$$

$$= \frac{1}{2}|\cos\theta\sin\phi - \sin\theta\cos\phi|^2 \qquad (18)$$

$$= \frac{1}{2}|\sin(\theta-\phi)|^2$$

The corresponding factor for photon 1 to pass and photon 2 not is

$$\frac{1}{2}\left|\langle\theta|0\rangle\left\langle\phi+\frac{\pi}{2}\left|\frac{\pi}{2}\right\rangle - \left\langle\theta\left|\frac{\pi}{2}\right\rangle\left\langle\phi+\frac{\pi}{2}\left|0\right\rangle\right|^2\right.\right.$$

$$= \frac{1}{2}|\cos(\theta-\phi)|^2 \qquad (19)$$

and so on. The probabilities for the various possible counting configurations are then

$$\rho(\text{yes, yes}) = \frac{x\Omega}{4\pi}\frac{1}{2}|\sin(\theta-\phi)|^2$$

$$\rho(\text{yes, no}) = \frac{x\Omega}{4\pi}\frac{1}{2}|\cos(\theta-\phi)|^2$$

$$\rho(\text{no, yes}) = \frac{x\Omega}{4\pi}\frac{1}{2}|\cos(\theta-\phi)|^2 \qquad (20)$$

$$\rho(\text{no, no}) = \frac{x\Omega}{4\pi}\frac{1}{2}|\sin(\theta-\phi)|^2 + x\left(1-\frac{\Omega}{4\pi}\right) + (1-x)$$

where x is the probability that the π^0 production mechanism actually works, Ω the (small) solid angle subtended by each counter at the production point, and no allowance has been made for bad timing, bad placing, or inefficient counting.

Now let us count $A = \pm 1$ for (yes/no) at 1 and $B = \pm 1$ for (yes/no) at 2.

Then the quantum mechanical mean value of the product is

$$P(\theta, \phi) = \rho(\text{yes}, \text{yes}) + \rho(\text{no}, \text{no}) - \rho(\text{yes}, \text{no}) - \rho(\text{no}, \text{yes})$$

$$= 1 - \frac{x\Omega}{4\pi}(1 + \cos 2(\theta - \phi)) \tag{21}$$

so that

$$|P(\theta, \phi) - P(\theta, \phi')| + P(\theta', \phi) + P(\theta', \phi') - 2$$

$$= \frac{x\Omega}{4\pi}\{|\cos 2(\theta - \phi) - \cos 2(\theta - \phi')| - \cos 2(\theta' - \phi) - \cos 2(\theta' - \phi') - 2\}$$

$$\tag{22}$$

The right-hand side of this expression is sometimes positive. Take in particular

$$\phi = 0, \quad 2\theta = \frac{\pi}{4}, \quad -2\phi' = \frac{\pi}{2}, \quad 2\theta' = \frac{3\pi}{4} \tag{23}$$

in which case the factor in curly brackets is

$$\{\quad\} = \frac{1}{\sqrt{2}} + \frac{1}{\sqrt{2}} + \frac{1}{\sqrt{2}} + \frac{1}{\sqrt{2}} - 2 = +2(\sqrt{2} - 1) \tag{24}$$

But if quantum mechanics were embeddable in a locally causal theory (16) would apply, with $a \to \theta$, $b \to \phi$, and c the implicit specification of the production mechanism, held fixed in (22). The right-hand side of (22) should then be *negative*. So quantum mechanics is *not* embeddable in a locally causal theory as formulated above.

6 Experiments

These considerations have inspired a number of experiments. The accuracy of quantum mechanics on the atomic scale makes it hard to believe that it could be seriously wrong on that scale in some hitherto undiscovered way. The ground state of the helium atom, for example, is just the kind of correlated wave function which is embarrassing, and its energy comes out right to very high accuracy. But perhaps it is sensible to verify that these curious correlations persist over macroscopic distances.

Experiments so far performed do not at all approach the ideal in which the settings of the instruments are determined only while the particles are in flight. When they are decided in advance, in space-time regions projecting into the overlap of the backward light cones, (16) does not follow from (12). For it was supposed in (12) that the complete specification n of the overlap is the same for the various cases compared. So one can imagine a theory which is locally causal in our sense but still manages to agree with quantum mechanics for static instruments. But it would have to contain a very clever mechanism by which the result registered by one instrument depends, after a suitable time lapse, on the setting of an arbitrarily distant instrument. So static experiments are also quite interesting.

Practical experiments are far removed from the ideal in other directions also. Geometrical and other inefficiencies lead to counters registering (no, no) with overwhelming probability, (yes, yes) very seldom, and (yes, no) and (no, yes) with probabilities only weakly dependent on the settings of the instruments. Then from (21)

$$P = 1 - \varepsilon^2$$

with ε^2 weakly dependent on the variables, so that (16) is trivially satisfied. The authors in general make some more or less *ad hoc* extrapolation to connect the results of the practical with the result of the ideal experiment. It is in this sense that the entirely unauthorized 'Bell's limit' sometimes plotted along with experimental points has to be understood. But such experiments also are of very high interest. For if quantum mechanics is to fail somewhere, and in the absence of a monstrous conspiracy, this should show up at some point on this side of the ideal gedanken experiment.

Several of these experiments[26] show impressive agreement with quantum mechanics, and exclude deviations as large as might be suggested by the locality inequality. Another experiment, very similar to one of those quoted[26], is said to be in agreement with it and yet in dramatic disagreement with quantum mechanics! And another experiment disagrees significantly with the quantum prediction. Of course any such disagreement, if confirmed, is of the utmost importance, and that independently of the kind of consideration we have been making here.

7 Messages

Suppose that we are finally obliged to accept the existence of these correlations at long range, and the gross non-locality of nature in the sense of this analysis. Can *we* then signal faster than light? To answer this we need at least a schematic theory of what *we* can do, a fragment of a theory of human beings. Suppose we can control variables like a and b above, but not those like A and B. I do not quite know what 'like' means here, but suppose that beables somehow fall into two classes, 'controllables' and 'uncontrollables'. The latter are no use for *sending* signals, but can be used for *reception*. Suppose that to A corresponds a quantum mechanical 'observable', an operator $\mathscr{A}$. Then if

$$\delta\mathscr{A}/\delta b \neq 0$$

we could signal between the corresponding space-time regions, using a change in b to induce a change in the expectation value of $\mathscr{A}$ or of some function of $\mathscr{A}$.

Suppose next that what we do when we change b is to change the quantum mechanical Hamiltonian $\mathscr{H}$ (say by changing some external field), so that

$$\delta \int \mathrm{d}t\, \mathscr{H} = \mathscr{B}\delta b$$

where $\mathscr{B}$ is again an 'observable' (i.e., an operator) localized in the region 2 of b. Then it is an exercise in quantum mechanics to show that if in a given reference system region (2) is entirely later in time than region (1)

$$\delta\mathscr{A}/\delta b = 0$$

while if the reverse is true

$$\delta\mathscr{A}/\delta b = [\mathscr{A}, -(1/\hbar)\mathscr{B}]$$

which is again zero (for spacelike separation) in quantum field theory by the usual local commutativity condition.

So if the ordinary quantum field theory is embedded in this way in a theory of beables, it implies that faster than light signalling is not possible. In this *human* sense relativistic quantum mechanics *is* locally causal.

8 Reservations and acknowledgements

Of course the assumptions leading to (16) can be challenged. Equation (22) may not embody *your* idea of local causality. You may feel that only the 'human' version of the last section is sensible and may see some way to make it more precise.

The space time structure has been taken as given here. How then about gravitation?

It has been assumed that the settings of instruments are in some sense free variables – say at the whim of experimenters – or in any case not determined in the overlap of the backward light cones. Indeed without such freedom I would not know how to formulate *any* idea of local causality, even the modest human one.

This paper has been an attempt to be rather explicit and general about the notion of locality, along lines only hinted at in previous publications

(Refs. 2, 4, 10, 19). As regards the literature on the subject, I am particularly conscious of having profited from the paper of Clauser, Horne, Holt and Shimony[3], which gave the prototype of (16), and from that of Clauser and Horne[16]. As well as a general analysis of the topic this last paper contains a valuable discussion of how best to apply the inequality in practice; I am indebted to it in particular for the point that in two-body decays (as compared with three-) the basic geometrical inefficiencies enter in (22) in a relatively harmless way. I have also profited from many discussions of the whole subject with Professor B. d'Espagnat.

References

1 N. Bohr, in *Albert Einstein*, Ed. Schilpp, Tudor (1).
2 J. S. Bell, *Physics* **1**, 195 (1965).
3 J. F. Clauser, R. A. Holt, M. A. Horne and A. Shimony, *Phys. Rev. Letters* **23**, 880 (1969).
4 J. S. Bell, in *Proceedings of the International School of Physics* Enrico Fermi, Course IL, Varenna 1970, Academic Press (1971).
5 R. Friedberg (1969, unpublished) referred to by M. Jammer[17].
6 E. P. Wigner, *Am. J. Phys.* **38**, 1005 (1970).
7 B. d'Espagnat, *Conceptual Foundations of Quantum Mechanics*, Benjamin (1971).
8 K. Popper, in *Perspectives in Quantum Theory*, Eds. W. Yourgrau and A. Van der Merwe, M.I.T. Press (1971).
9 H. P. Stapp, *Phys. Rev.* **D3**. 1303 (1971).
10 J. S. Bell, *Science* **177**, 880 (1972).
11 P. M. Pearle, *Phys. Rev.* **D2**, 1418 (1970).
12 J. H. McGuire and E. S. Fry, *Phys. Rev.* **D7**, 555 (1972).
13 S. Freedman and E. P. Wigner, *Foundations of Physics* **3**, 457 (1973).
14 F. J. Belinfante, *A Survey of Hidden Variable Theories*, Pergamon (1973).
15 V. Capasso, D. Fortunato and F. Selleri, *Int. J. Theor. Phys.* **7**, 319 (1973).
16 J. F. Clauser and M. A. Horne, *Phys. Rev*, **D10**, 526 (1974).
17 M. Jammer, *The Philosophy of Quantum Mechanics*, Wiley (1974). See in particular references to T. D. Lee (p. 308) and R. Friedberg (pp. 244, 309, 324).
18 D. Gutkowski and G. Masotto, *Nuovo Cimento* **22B**, 1921 (1974).
19 J. S. Bell, in *The Physicist's Conception of Nature*, Ed. J. Mehra and D. Reidel (1973).
20 B. d'Espagnat, *Phys. Rev.* **D11**, 1424 (1975).
21 G. Corleo, D. Gutkowski, G. Masotto, and M. V. Valdes, *Nuovo Cimento*, **B25**, 413–24 (1975).
22 H. P. Stapp, *Nuovo Cimento*, **B29**, 270–6 (1975).
23 D. Bohm and B. Hiley, *Foundations of Physics*, **5**, 93–109 (1975).
24 A. Baracca, S. Bergia and M. Restignoli. *Conference on Few Body Problems, Quebec, Aug.1974*, 68–9. Quebec, Laval University Press (1975).
25 A. Baracca, D. J. Bohm, R. J. Hiley and A. E. G. Stuart, *Nuovo Cimento*, **28B**, 453–66 (1975).
26 For a short review of experiments see paper 10 in this collection.

8

Locality in quantum mechanics: reply to critics

The editor has asked me to reply to a paper, by G. Lochak[1], refuting a theorem of mine on hidden variables. If I understand correctly, Lochak finds that I failed somehow to allow for the effect on these variables of the measuring equipment. I will try to explain why I do not agree. The opportunity will also be taken here to comment on another refutation[2], by L. de la Peña, A. M. Cetto and T. A. Brody, and on another[3] by L. de Broglie. Yet another refutation of the same theorem, by J. Bub[4], has already been refuted by S. Freedman and E. P. Wigner[5].

Let us recall a typical context to which the theorem is relevant. A 'pair of spin $\frac{1}{2}$ particles' is produced in a space-time region 3 and activates counting systems, preceded by Stern–Gerlach magnets, in space–time regions 1 and 2. The system at 1 is such that one of two counters ('up' or 'down') registers each time the experiment is done; correspondingly we label the result there by A ($= +1$ or -1). Likewise the system at 2 is such that one of two counters registers each time the experiment is done, giving B ($= +1$ or -1). We are interested in correlations between the counts in 1 and 2, and define a correlation function

$$\overline{AB}$$

which is the average of the product of A and B over many repetitions of the experiment.

Now it would certainly be better to give a purely operational, technological, macroscopic, description of the equipment involved. This would avoid completely any use of words like 'particle' and 'spin', and so avoid the possibility that someone feels obliged to form a personal microscopic picture of what is going on. But it would take quite long to give such a purely technological specification. So, please accept that the words 'particle' and 'spin' are used here only as part of a conventional shorthand, to invoke without lengthy explicit description the *kind* of experimental equipment involved, and with no commitment whatever to any picture of what, if anything, really causes the counters to count.

Suppose that part of the specification of the equipment is by two unit vectors $\hat{\mathbf{a}}$ and $\hat{\mathbf{b}}$ (e.g., the directions of certain magnetic fields at 1 and 2). Then according to ordinary quantum mechanics situations exist for which

$$\overline{AB} = -\hat{\mathbf{a}}\cdot\hat{\mathbf{b}} \tag{1}$$

to good accuracy.

Actually it is this last statement which is challenged by de Broglie. Although his paper is called 'Sur la réfutation du théorème de Bell', it is not in fact concerned with any reasoning of mine. He is of the opinion that the correlation function (1) simply cannot occur for macroscopic separations, either in nature or in ordinary quantum mechanics: 'Nous échappons complètement à cette objection puisque, pour nous, les mesures du spin sur des électrons éloignés ne sont pas corrélées'. As regards ordinary quantum mechanics, de Broglie disagrees here with most students of the subject, and I am unable to follow his reasons for doing so. As regards nature, he seems to disagree also with experiment[6].

Now we investigate the hypothesis that the final state of the system, in particular A and B, would be fully determined by the equations of some theory if the initial conditions were fully specified. So to parameters like $\hat{\mathbf{a}}$ and $\hat{\mathbf{b}}$, subject to experimental manipulation, we add a list of hypothetical 'hidden' parameters λ. We can take these λ to be the *initial* values (say just after the action of the source) of some corresponding dynamical variables. We have no interest in what subsequently happens to these variables except in so far as they enter into the measurement results A and B. But in so far as they do enter into A and B we allow *fully for the effect of the measuring equipment by allowing A and B to depend not only on the initial values λ of the hidden parameters but also on the parameters $\hat{\mathbf{a}}$ and $\hat{\mathbf{b}}$, specifying the measuring devices*:

$$A(\hat{\mathbf{a}}, \hat{\mathbf{b}}, \lambda)$$
$$B(\hat{\mathbf{a}}, \hat{\mathbf{b}}, \lambda) \tag{2}$$

We have no need to enquire into the precise nature of this dependence on $\hat{\mathbf{a}}$ and $\hat{\mathbf{b}}$, nor into how it comes about, *whether by the effect of the measuring equipment on the hidden variables* of which the λ are the *initial* values, or otherwise.

Can one find some functions (2) and some probability distribution $\rho(\lambda)$ which reproduces the correlation (1)? Yes, many, but now we add the hypothesis of *locality*, that the setting $\hat{\mathbf{b}}$ of a particular instrument has no effect on what happens, A, in a remote region, and likewise that $\hat{\mathbf{a}}$ has no effect on B:

$$A(\hat{\mathbf{a}}, \lambda)$$
$$B(\hat{\mathbf{b}}, \lambda) \tag{3}$$

With these *local* forms, it is *not* possible to find functions A and B and a probability distribution ρ which give the correlation (1). This is the theorem. The proof will not be repeated here.

Lochak illustrates the way in which the output of a single instrument A depends on its setting $\hat{\mathbf{a}}$, as allowed for in (3), in the hidden parameter theory of de Broglie. I think this is very instructive. But more instructive for the present purpose is the case of *two* instruments and *two* particles. *Then one finds that in de Broglie's theory the dependence is not of the local form (3) but of the nonlocal form (2).* I have made this point on several occasions, in two of the three papers referred to by Lochak and elsewhere[7]. It may be that Lochak has in mind some other extension of de Broglie's theory, to the more-than-one-particle system, than the straightforward generalization from 3 to $3N$ dimensions that I considered. But if his extension is local it will not agree with quantum mechanics, and if it agrees with quantum mechanics it will not be local. This is what the theorem says.

The objection of de la Peña, Cetto, and Brody is based on a misinterpretation of the demonstration of the theorem. In the course of it reference is made to

$$A(\hat{\mathbf{a}}', \lambda) \quad , \quad B(\hat{\mathbf{b}}', \lambda)$$

as well as

$$A(\hat{\mathbf{a}}, \lambda) \quad , \quad B(\hat{\mathbf{b}}, \lambda)$$

These authors say 'Clearly, since A, A', B, B' are all evaluated for the same λ, they must refer to four measurements carried out on the same electron–positron pair. We can suppose, for instance, that A' is obtained after A, and B' after B'. But by no means. We are not at all concerned with sequences of measurements on a given particle, or of pairs of measurements on a given pair of particles. We are concerned with experiments in which for each pair the 'spin' of each particle is measured once only. The quantities

$$A(\hat{\mathbf{a}}', \lambda) \quad , \quad B(\hat{\mathbf{b}}', \lambda)$$

are just the same functions

$$A(\hat{\mathbf{a}}, \lambda) \quad , \quad B(\hat{\mathbf{b}}, \lambda)$$

with different arguments.

References

1 G. Lochak, *Fundamenta Scientiae* (Université de Strasbourg, 1975), No 38, reprinted in *Epistemological Letters*, p. 41, September 1975.
2 L. de la Pena, A. M. Cetto and T. A. Brody, *Nuovo Cimento Letters* **5**, 177 (1972).
3 L. de Broglie, *CR Acad. Sci. Paris* **278**, B721 (1974).
4 J. Bub, *Found. Phys.* **3**, 29 (1973).
5 S. Freedman and E. Wigner, *Found. Phys.* **3**, 457 (1973).
6 S. J. Freedman and J. F. Clauser, *Phys. Rev. Lett.* **28**, 938 (1972). A brief account is given by M. Paty, *Epistemological Letters*, p. 31, September 1975.
7 J. S. Bell, *On the Hypothesis that the Schrödinger Equation is Exact*, CERN Preprint TH. 1424 (1971).

9

How to teach special relativity

I have for long thought that if I had the opportunity to teach this subject, I would emphasize the continuity with earlier ideas. Usually it is the discontinuity which is stressed, the radical break with more primitive notions of space and time. Often the result is to destroy completely the confidence of the student in perfectly sound and useful concepts already acquired[1].

If you doubt this, then you might try the experiment of confronting your students with the following situation[2]. Three small spaceships, A, B, and C, drift freely in a region of space remote from other matter, without rotation and without relative motion, with B and C equidistant from A (Fig. 1).

On reception of a signal from A the motors of B and C are ignited and they accelerate gently[3] (Fig. 2).

Let ships B and C be identical, and have identical acceleration programmes. Then (as reckoned by an observer in A) they will have at every moment the same velocity, and so remain displaced one from the other by a fixed distance. Suppose that a fragile thread is tied initially between projections from B and C (Fig. 3). If it is just long enough to span the required distance initially, then as the rockets speed up, it will become too short, because of its need to Fitzgerald contract, and must finally break. It must break when, at a sufficiently high velocity, the artificial prevention of the natural contraction imposes intolerable stress.

Fig. 1.

Is it really so? This old problem came up for discussion once in the CERN canteen. A distinguished experimental physicist refused to accept that the thread would break, and regarded my assertion, that indeed it would, as a personal misinterpretation of special relativity. We decided to appeal to the CERN Theory Division for arbitration, and made a (not very systematic) canvas of opinion in it. There emerged a clear consensus that the thread would **not** break!

Of course many people who give this wrong answer at first get the right answer on further reflection. Usually they feel obliged to work out how things look to observers B or C. They find that B, for example, sees C drifting further and further behind, so that a given piece of thread can no longer span the distance. It is only after working this out, and perhaps only with a residual feeling of unease, that such people finally accept a conclusion which is perfectly trivial in terms of A's account of things, including the Fitzgerald contraction. It is my impression that those with a more classical education, knowing something of the reasoning of Larmor, Lorentz, and Poincaré, as well as that of Einstein, have stronger and sounder instincts. I will try to sketch here a simplified version of the Larmor–Lorentz–Poincaré approach that some students might find helpful.

Some familiarity with Maxwell's equations is assumed, so that the calculation of the field of a moving point charge can be followed, or at least the result accepted without mystification. For a charge Ze moving with constant velocity V along the z axis the nonvanishing field components are:

$$\left.\begin{aligned}
E_z &= Zez'(x^2 + y^2 + z'^2)^{-3/2} \\
E_x &= Zex(x^2 + y^2 + z'^2)^{-3/2}(1 - V^2/c^2)^{-1/2} \\
E_y &= Zey(x^2 + y^2 + z'^2)^{-3/2}(1 - V^2/c^2)^{-1/2} \\
B_x &= -(V/c)E_y \\
B_y &= +(V/c)E_x
\end{aligned}\right\} \tag{1}$$

Fig. 2.

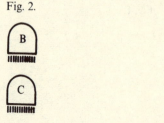

Fig. 3.

where

$$z' = (z - z_N(t))(1 - V^2/c^2)^{-1/2} \qquad (2)$$

and $z_N(t)$ is the position of the charge at time t. For a charge at rest, $V = 0$, this is just the familiar Coulomb field, spherically symmetrical about the source. But when the source moves very quickly, so that V^2/c^2 is not very small, the field is no longer spherically symmetrical. The magnetic field is transverse to the direction of motion and, roughly speaking, the system of lines of electric field is flattened in the direction of motion (Fig. 4).

In so far as microscopic electrical forces are important in the structure of matter, this systematic distortion of the field of fast particles will alter the internal equilibrium of fast moving material. It is to be expected therefore that a body set in rapid motion will change shape. Such a change of shape, the Fitzgerald contraction, was in fact postulated on empirical grounds by G. F. Fitzgerald in 1889 to explain the results of certain optical experiments.

Fig. 4.

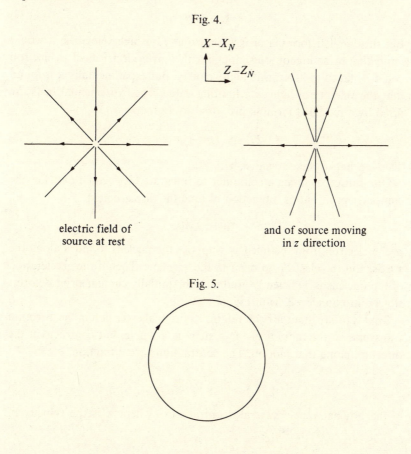

electric field of
source at rest

and of source moving
in z direction

Fig. 5.

The simplest piece of matter that we can discuss in this connection is a single atom. In the classical model of such an atom a number of electrons orbit around a nucleus. For simplicity take only one electron, and ignore the effect, on the relatively masssive nucleus, of the field of the electron. The dynamical problem is then that of the motion of the electron in the field of the nucleus. Let us start with the nucleus at rest and the electron, for simplicity, describing a circular orbit (Fig. 5).

What happens to this orbit when the nucleus is set in motion?[4]

If the acceleration of the nucleus is quite gentle, its field differs only slightly from (1). Moreover, the accurate expression is known[5].

In this field we have to solve the equation of motion for the electrons

$$\frac{d\mathbf{p}}{dt} = - e(\mathbf{E} + c^{-1}\dot{\mathbf{r}}_e \times \mathbf{B}) \tag{3}$$

where $\mathbf{r}_e$ is the electron position and the fields in (3) are evaluated at that position. At low velocity, momentum and velocity are related by

$$\dot{\mathbf{r}}_e = \mathbf{p}/m \tag{4}$$

But this familiar formula proves inadequate for high velocities. It would imply that by acting for long enough with a given electric field an electron could be taken to arbitrarily high velocity. But experimentally it is found that the velocity of light is a limiting value. The experimental facts are fitted by a modified formula proposed by Lorentz

$$\dot{\mathbf{r}}_e = \mathbf{p}/\sqrt{m^2 + \mathbf{p}^2 c^{-2}} \tag{5}$$

This is what we take together with (3).

One can programme a computer to integrate these equations. Let the computer print out as a function of time the displacement

$$\mathbf{r}_e(t) - \mathbf{r}_N(t)$$

of the electron from the nucleus. Suppose the nucleus to move along the z axis, and the electron to orbit in the xz plane. Then if the acceleration of the nucleus is sufficiently gradual[6], the initially circular orbit deforms slowly into an ellipse, as in Fig. 6.

That is to say that the orbit retains its original extension in the direction transverse to the motion of the system as a whole, but contracts in the direction along that motion. The contraction is to a fraction

$$\sqrt{1 - V^2/c^2} \tag{6}$$

of the original – the Fitzgerald contraction – where V is the velocity of

Fig. 6.

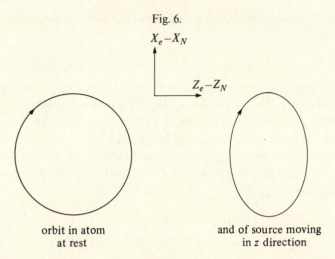

orbit in atom   and of source moving
at rest   in z direction

the nucleus during the orbit in question. Moreover, this is performed in a period exceeding the original period by a factor

$$1/\sqrt{1 - V^2/c^2} \tag{7}$$

– the time-dilation of J. Larmor (1900).

If the period of the system at rest is T, then the total number of revolutions during a journey of time t with proton velocity $V(t)$ is

$$T^{-1} \int_0^t d\tau \sqrt{1 - c^{-2} V(\tau)^2} \tag{8}$$

– which is less than that for a similar system at rest, even if the moving system is both initially and finally also at rest and initially and finally in the same position. This straightforward result of computation is the origin of the 'paradox' of the travelling twin (Le Voyageur de Langevin, en français).

These results suggests that it may be useful to describe the moving system in terms of new variables which incorporate the Fitzgerald and Larmor effects:

$$\left. \begin{aligned} z' &= (z - z_N(t))/\sqrt{1 - c^{-2} V(t)^2} \\ x' &= x \quad y' = y \\ t' &= \int_0^t d\tau \sqrt{1 - c^{-2} V(t)^2} - c^{-2} V(t) Z' \end{aligned} \right\} \tag{9}$$

The motivation for the last term in the definition of t' is not obvious, but emerges from more detailed examination of the orbit. Including this term, the orbit

$$z'_e(t') \quad , \quad x'_e(t') \tag{10}$$

is not merely circular, with period T, but is swept out with constant angular velocity. That is, *the description of the orbit of the moving atom in terms of the primed variables is identical with the description of the orbit of the stationary atom in terms of the original variables.*

As regards the electromagnetic field we have already profited from the use of the variable z' in writing (1). Going further in this direction, one can introduce

$$\left. \begin{aligned}
E'_x &= (E_x - c^{-1}VB_y)/\sqrt{1 - c^{-2}V^2} \\
E'_y &= (E_y + c^{-1}VB_x)/\sqrt{1 - c^{-2}V^2} \\
E'_z &= E_z \\
B'_x &= (B_x + c^{-1}VE_y)/\sqrt{1 - c^{-2}V^2} \\
B'_y &= (B_y - c^{-1}VE_x)/\sqrt{1 - c^{-2}V^2} \\
B'_z &= B_z
\end{aligned} \right\} \tag{11}$$

Then it is easy to check that *the expression of the field of the uniformly moving charge in terms of the primed variables is identical with the expression of the field of the stationary charge in terms of the original variables.*

We have been speaking of a *gently* accelerated atom. So the velocity V always remains essentially constant during many revolutions of the electron. During any such interval, one can arrange that

$$\int_0^t d\tau \sqrt{1 - c^{-2}V(\tau)^2} = t\sqrt{1 - c^{-2}V^2} \tag{12}$$

$$z_N(t) = Vt \tag{13}$$

by a suitable choice of the origin of z and t. Then (9) can be rewritten

$$\left. \begin{aligned}
z' &= (z - Vt)/\sqrt{1 - V^2/c^2} \\
x' &= x \\
y' &= y \\
t' &= (t - Vx/c^2)/\sqrt{1 - V^2/c^2}
\end{aligned} \right\} \tag{14}$$

This is then the standard form of what is called a *Lorentz transformation.* That the use of such variables enables the moving atom to be described by the functions appropriate to the stationary atom is an illustration of the following exact mathematical fact. When Maxwell's equations

$$\frac{1}{c}\frac{\partial E_x}{\partial t} = \frac{\partial B_z}{\partial y} - \frac{\partial B_y}{\partial z}, \text{ etc.} \tag{15}$$

and the Lorentz equations

$$\left. \begin{aligned} \frac{d\mathbf{p}}{dt} &= -e(\mathbf{E} + c^{-1}\mathbf{r}_e \times \mathbf{B}) \\[2ex] \frac{d\mathbf{r}_e}{dt} &= \mathbf{p}/\sqrt{m^2 + c^{-2}\mathbf{p}^2} \end{aligned} \right\} \tag{16}$$

are expressed in terms of the new variables (11) and (14) *they have exactly the same form as before*

$$\left. \begin{aligned} \frac{1}{c}\frac{\partial E'_x}{\partial t'} &= \frac{\partial B'_z}{\partial y'} - \frac{\partial B'_y}{\partial z'}, \text{ etc.} \\[2ex] \frac{d\mathbf{p}'}{dt'} &= -e(\mathbf{E}' + c^{-1}\mathbf{r}'_e \times \mathbf{B}') \\[2ex] \frac{d\mathbf{r}'_e}{dt'} &= \mathbf{p}'/\sqrt{m^2 + c^{-2}\mathbf{p}'^2} \end{aligned} \right\} \tag{17}$$

(where the last equation can be taken as defining $\mathbf{p}'$). The equations are said to be *Lorentz invariant*. From any solution of the original equations, involving certain mathematical functions (e.g., the Coulomb field and the circular orbit in the stationary atom), one can construct a new solution by putting primes on all the variables and then eliminating these primes by means of (11) and (14) (giving, e.g., the flattened field and ellipitcal orbit of the moving atom). Moreover, by a trivial extension this reasoning applies not only to a single electron interacting with a single electro-magnetic field, but to any number of charged particles, each interacting with the fields of all others. This allows an extension to very complicated systems of some of the results described above for the simple atom. Given any state of the complicated system, there is a corresponding 'primed' state which is in overall motion with respect to the original, shows the Fitzgerald contraction, and the Larmor dilation. Suppose, for example, in the original state all particles are permanently inside a region bounded by

$$z = \pm L/2$$

then the corresponding primed state has boundaries

$$z' = \pm L/2$$

or from (14)

$$z = Vt \pm 1/2L\sqrt{L - V^2/c^2}$$

i.e., they move with the velocity V and are closer together by the Fitzgerald factor.

Suppose next that in the original state something happens (e.g., an electron passes) at a place $x = x_1, y = y_1, z = z_1$ at time t_1, and again at the same place at time t_2. Then the corresponding events in the primed state occur at

$$x' = x_1 \quad y' = y_1 \quad z' = z_1 \quad t' = t_1, t_2$$

or (solving (14)) at

$$x = x_1 \quad y = y_1$$

$$z = \frac{z_1 + Vt_1}{\sqrt{1 - V^2/c^2}}, \frac{z_1 + Vt_2}{\sqrt{1 - V^2/c^2}}$$

$$t = \frac{t_1 + Vz_1/c^2}{\sqrt{1 - V^2/c^2}}, \frac{t_2 + Vz_1/c^2}{\sqrt{1 - V^2/c^2}}$$

The place of occurrence moves with velocity V, and the time interval between the two events increases by the Larmor factor.

Can we conclude then that an arbitrary system, set in motion, will show precisely the Fitzgerald and Larmor effects? Not quite. There are two provisos to be made.

The first is this: the Maxwell–Lorentz theory provides a very inadequate model of actual matter, in particular solid matter. It is not possible in a classical model to reproduce the empirical stability of such matter. Moreover, things are made worse when radiation reaction is included. Moving charges in general radiate energy and momentum, and because of this there are extra small terms in the equation of motion. Even in the simple hydrogen atom the electron then spirals in towards the proton instead of remaining in a stable orbit. These problems were among those which led to the replacement of classical by quantum theory. Moreover, even in the quantum theory electromagnetic interactions turn out to be not the only ones. For example, atomic nuclei are apparently held together by quite different 'strong' interactions. We do not need to get involved in these details if we assume with Lorentz that the *complete theory* is Lorentz invariant, in that the equations are unchanged by the change of variables (14), supplemented by some generalization of (13) to cover all the quantities in the theory. Then for any state there is again a corresponding primed state, showing the Fitzgerald and Larmor effects.

The second proviso is this. Lorentz invariance alone shows that for any state of a system at rest there is a corresponding 'primed' state of that system in motion. But it does not tell us that if the system is set anyhow in motion, it will actually go into the 'prime' of the original state, rather than into the 'prime' of some *other* state of the system at rest. In fact, it will generally do the latter. A system set brutally in motion may be bruised, or broken, or heated, or burned. For the simple classical atom similar things could have happened if the nucleus, instead of being moved smoothly, had been *jerked*. The electron could be left behind completely. Moreover, a given acceleration is or is not sufficiently gentle depending on the orbit in question. An electron in a small, high frequency, tightly bound orbit, can follow closely a nucleus that an electron in a more remote orbit – or in another atom – would not follow at all. Thus we can only assume the Fitzgerald contraction, etc., for a coherent dynamical system whose configuration is determined essentially by internal forces and only little perturbed by gentle external forces accelerating the system as a whole. Let us do so.

Then, for example, in the rocket problem of the introduction, the material of the rockets, and of the thread, will Lorentz contract. A sufficiently strong thread would pull the rockets together and impose Fitzgerald contraction on the combined system. But if the rockets are too massive to be appreciably accelerated by the fragile thread, the latter has to break when the velocity becomes sufficiently great.

So far we have discussed moving *objects*, but not yet moving *subjects*. The question of moving observers is not entirely academic. Quite apart from people in rockets, it seems reasonable to regard the earth itself, orbiting the sun, as moving – at least for much of the year[7]. The important point to be made about moving observers is this, given Lorentz invariance: *the primed variables, introduced above simply for mathematical convenience, are precisely those which would naturally be adopted by an observer moving with constant velocity who imagines herself to be at rest.* Moreover, such an observer will find that the laws of physics in these terms are precisely those that she learned when at rest (if she was taught correctly).

Such an observer will naturally take for the origin of space coordinates a point at rest with respect to herself. This accounts for the Vt term in the relation

$$z' = (z - Vt)/\sqrt{1 - V^2/c^2}$$

The factor $\sqrt{1 - V^2/c^2}$ is accounted for by the Fitzgerald contraction of her metre sticks. But will she not *see* that her metre sticks are contracted when laid out in the z direction – and even decontract when turned in the x direction? No, because the retina of her eye will also be contracted, so that just the same cells receive the image of the metre stick as if both stick and observer were at rest. In the same way she will not notice that her clocks have slowed down, because she will herself be thinking more slowly. Moreover, imagining herself to be at rest, she will not know that light overtakes her, or comes to meet her, with different relative velocities $c \pm v$. This will mislead her in synchronizing clocks at different places, so that she is led to think that

$$t' = \frac{t - Vz/c^2}{\sqrt{1 - V^2/c^2}}$$

is the real time, for with this choice light again *seems* to go with velocity c in all directions. This can be checked directly, and is also a consequence of the prime Maxwell equations. In measuring electric field she will use a test charge at rest with respect to her equipment, and so measure actually a combination of **E** and **B**. Defining both **E** and **B** by requiring what looks like the familiar effects on moving charged particles, she will be led rather to **E'** and **B'**. Then she will be able to verify that all the laws of physics are as she remembers, at the same time confirming her own good sense in the definitions and procedures that she has adopted. If something does not come out right, she will find that her apparatus is in error (perhaps damaged during acceleration) and repair it.

Our moving observer O', imagining herself to be at rest, will imagine that it is the stationary observer O who moves. And it is as easy to express his variables in terms of hers as vice versa

$$\left. \begin{aligned} x' = x \quad y' = y \\ z' = \frac{z - Vt}{\sqrt{1 - V^2/c^2}} \\ t' = \frac{t - Vz/c^2}{\sqrt{1 - V^2/c^2}} \end{aligned} \right\} \Leftrightarrow \left\{ \begin{aligned} x = x' \quad y = y' \\ z = \frac{z' + Vt'}{\sqrt{1 - V^2/c^2}} \\ t = \frac{t' + Vz'/c^2}{\sqrt{1 - V^2/c^2}} \end{aligned} \right.$$

Only the sign of *V* changes. She will say that *his* metre sticks have contracted, that *his* clocks run slow, and that *he* has not synchronized properly clocks at different places. She will attribute his use of wrong variables to these Fitzgerald–Larmor–Lorentz–Poincaré effects in *his* equipment. Her view will be logically consistent and in perfect accord with the observable facts. He will have no way of persuading her that she is wrong.

This completes the introduction to what has come to be called 'the special theory of relativity'. It arose from experimental failure to detect any change, in the apparent laws of physics in terrestrial laboratories, with the slowly changing orbital velocity of the earth. Of particular importance was the Michelson–Morley experiment, which attempted to find some difference in the apparent velocity of light in different directions.

We have followed here very much the approach of H. A. Lorentz. Assuming physical laws in terms of certain variables (t, x, y, z), an investigation is made of how things look to observers who, with their equipment, in terms of these variables, move. It is found that if physical laws are Lorentz invariant, such moving observers will be unable to detect their motion. As a result it is not possible experimentally to determine which, if either, of two uniformly moving systems, is really at rest, and which moving. All this for *uniform* motion: accelerated observers are not considered in the 'special' theory.

The approach of Einstein differs from that of Lorentz in two major ways. There is a difference of philosophy, and a difference of style.

The difference of philosophy is this. Since it is experimentally impossible to say which of two uniformly moving systems is *really* at rest, Einstein declares the notions 'really resting' and 'really moving' as meaningless. For him only the *relative* motion of two or more uniformly moving objects is real. Lorentz, on the other hand, preferred the view that there is indeed a state of *real* rest, defined by the 'aether', even though the laws of physics conspire to prevent us identifying it experimentally. The facts of physics do not oblige us to accept one philosophy rather than the other. And we need not accept Lorentz's philosophy to accept a Lorentzian pedagogy. Its special merit is to drive home the lesson that the laws of physics in any *one* reference frame account for all physical phenomena, including the observations of moving observers. And it is often simpler to work in a single frame, rather than to hurry after each moving object in turn.

The difference of style is that instead of inferring the experience of moving observers from known and conjectured laws of physics, Einstein starts from the *hypothesis* that the laws will look the same to all observers in uniform motion. This permits a very concise and elegant formulation of the theory, as often happens when one big assumption can be made to cover several less big ones. There is no intention here to make any reservation whatever about the power and precision of Einstein's approach. But in my opinion there is also something to be said for taking students along the road made by Fitzgerald, Larmor, Lorentz and Poincaré[8]. The longer road sometimes gives more familiarity with the country.

In connection with this paper I warmly acknowledge the counsels of

M. Bell, F. Farley, S. Kolbig, H. Wind, A. Zichichi and H. Øveras. I thank especially H. D. Deas for discussion of these ideas at an early stage.

Notes and references

1 Notes are to be ignored in a first reading.
2 E. Dewan & M. Beran, *Am. J. Phys.* **27**, 517, 1959. A. A. Evett & R. K. Wangsness, *Am. J. Phys.* **28**, 566, 1960. E. M. Dewan, *Am. J. Phys.* **31**, 383, 1963. A. A. Evett, *Am. J. Phys.* **40**, 1170, 1972.
3 Violent acceleration could break the thread just because of its own inertia while velocities are still small. This is not the effect of interest here. With gentle acceleration the breakage occurs when a certain *velocity* is reached, a function of the degree to which the thread permits stretching beyond its natural length.
4 This method of acceleration, applying somehow a force to the nucleus without any direct effect on the electron, is not very realistic. However, as explained later, it follows from Lorentz invariance and stability considerations that any sufficiently smooth acceleration process will produce the same Fitzgerald contraction and Larmor dilation. The student is invited to attach a meaning to this statement also in the more general cases of non-circular orbits and when the acceleration is not in the plane of the orbit.
5 For a source of charge Ze the fields are[9], in c.g.s. units,

$$\mathbf{E} = \frac{Ze}{s^3}\left\{\left(\mathbf{r} - r\frac{[\mathbf{v}]}{c}\right)\left(1 - \frac{[\mathbf{v}]^2}{c^2}\right) + \left(\left(\mathbf{r} - r\frac{[\mathbf{v}]}{c}\right) \times \frac{[\mathbf{A}]}{c^2}\right)\right\} \qquad (5.1)$$

$$\mathbf{B} = \mathbf{r} \times \mathbf{E}/r$$

where
$$\mathbf{r} = \mathbf{r}_e - [\mathbf{r}_N]$$
$$s = r - \mathbf{r}\cdot[\mathbf{v}]/c.$$

These are the fields at position $\mathbf{r}_e$ at time t due to a source which *at the retarded time*

$$t - r/c \qquad (5.2)$$

had position, velocity, and acceleration

$$[\mathbf{r}_N], [\mathbf{v}], [\mathbf{A}].$$

Because of the appearance of r in the retarded time (5.2), which is itself needed to calculate $\mathbf{r}$, these equations are less explicit than could be desired.

However, if one starts with a situation in which the source has been at rest for some time, r is initially just the instantaneous distance to the source. One can keep track of it subsequently by integrating the differential equation

$$\frac{dr}{dt} = s^{-1}\mathbf{r}\cdot(\dot{\mathbf{r}}_e - [\mathbf{v}]) \qquad (5.3)$$

which follows from

$$\mathbf{r}^2 = (\mathbf{r}_e - [\mathbf{r}_N])\cdot(\mathbf{r}_e - [\mathbf{r}_N])$$

on differentiating with respect to time, noting that

$$\frac{d}{dt}[\mathbf{r}_N] = [\mathbf{v}]\left(1 - \frac{dr}{cdt}\right)$$

In the particular case of uniform motion, $A = 0$, the retarded quantities can be expressed in terms of unretarded ones:

$$\left.\begin{aligned}
[\mathbf{A}] &= \mathbf{A} = 0 \\
[\mathbf{v}] &= \mathbf{v} = \text{constant} \\
[\mathbf{r}_N] &= \mathbf{r}_N - \mathbf{v}r/c \\
r &= \frac{c^{-1}\mathbf{v}\cdot(\mathbf{r}_e - \mathbf{r}_N) + \sqrt{(c^{-1}\mathbf{v}\cdot(\mathbf{r}_e - \mathbf{r}_N))^2 + (\mathbf{r}_e - \mathbf{r}_N)^2(1 - v^2/c^2)}}{(1 - v^2/c^2)}
\end{aligned}\right\} \quad (5.4)$$

the last expression being the solution of

$$r^2 = (\mathbf{r}_e - \mathbf{r}_N + c^{-1}r\mathbf{v})^2$$

With these expressions (5.1) reduces to (1).

6 To verify this for the hydrogen atom ($Z = 1$) with a realistic orbit radius, e.g., the Bohr radius

$$h(mcZ\alpha)^{-1}\sqrt{1 - (Z\alpha)^2}$$

where α is the fine structure constant, $\sim 1/137$, might require much computing time. The acceleration has to be very gentle, because the internal forces are weak, and because the orbit is close to an 'integral resonance instability' (in the language of particle accelerator theory). Taking a larger value of Z, e.g. $Z \sim 70$, much larger accelerations are possible and a modest computing time suffices. The idea of obtaining the Fitzgerald and Larmor effects in such a system, by straight-forward integration of equations of motion, was perhaps suggested to me by a remark of J. Larmor[10].

7 Conceivably the motion of the earth relative to the sun, and the motion of the sun itself relative to whatever inertial frame we adopt, could conspire to make the earth itself momentarily at rest. But this situation would not persist as the earth continues round the sun, assuming the latter to move rather uniformly. By the way, the orbital velocity of the earth is about 3×10^5 cm/sec. The velocity of the earth's surface relative to the centre, due to the daily rotation, is about one hundredth of this.

8 The only modern text-book taking essentially this road, among those with which I am acquainted, seems to be that of L. Janossy: *Theory of Relativity Based on Physical Reality*, Académiaia Kiado, Budapest (1971).

9 These fields follow from the point-source retarded potentials of Lienard (1898) and Wiechert (1900). See, for example, W. K. H. Panofsky and M. Phillips: *Classical Electricity and Magnetism*. Addison-Wesley (1964), Eqs. 20-13, 20-15.
Unfortunately, for our purpose, in modern textbooks this material is usually presented after chapters on relativity. But the incidental reference to relativity, which can then appear, can be disregarded; the business at hand is just the writing down of certain solutions of Maxwell's equations.

10 J. Larmor, *Aether and Matter*. Cambridge (1900) p. 179. The example is used by Larmor to illustrate a very general correspondence between stationary and moving systems, based on what is now called the Lorentz invariance of the Maxwell equations, which Larmor establishes to second order in v/c. Note that he does not write separate equations for the motion of sources, like our (3) and (5). He seems to have in mind a model in which the motion of singularities is dictated somehow by the field equations, in analogy with the motion of vortex lines in hydrodynamics. Larmor summarizes his general conclusions on p. 176:
'We derive the result, correct to the second order, that if the internal forces of a

material system arise wholly from electrodynamic actions between the systems of electrons which constitute the atoms, then an effect of imparting to a steady material system a uniform velocity of translation is to produce a uniform contraction of the system in the direction of the motion, of amount $\varepsilon^{-1/2}$ or $1 - 1/2v^2/C^2$. The electrons will occupy corresponding positions in this contracted system, but the aethereal displacements in the space around them will not correspond: if (f, g, h) and (a, b, c) are those of the moving system, then the electric and magnetic displacements at corresponding points of the fixed systems will be the values that the vectors

$$\varepsilon^{1/2}\left(\varepsilon^{-1/2}f, g - \frac{v}{4\pi G^2}c, h + \frac{v}{4\pi C^2}b \right)$$

and

$$\varepsilon^{1/2}(\varepsilon^{-1/2}a, b + 4\pi vh, c - 4\pi vg)$$

had at a time const. $+ vx/C^2$ before the instant considered when the scale of time is enlarged in the ratio $\varepsilon^{1/2}$.

The special example is described on p. 179:

'As a simple illustration of the general molecular theory, let us consider the group formed of a pair of electrons of opposite signs describing steady circular orbits round each other in a position of rest. (The orbital velocities are in this illustration supposed so small that radiation is not important): we can assert from the correlation, that when this pair is moving through the aether with velocity v in a direction lying in the plane of their orbits, these orbits relative to the translatory motion will be flattened along the direction of v to ellipticity $1 - 1/2v^2/C^2$, while there will be a first-order retardation of phase in each orbital motion when the electron is in front of the mean position combined with acceleration when behind it so that on the whole the period will be changed only in the second-order ratio $1 + 1/2v^2/C^2$. The specification of the orbital modification produced by the translatory motion, for the general case when the direction of that motion is inclined to the plane of the orbit, may be made similarly: it can also be extended to an ideal molecule constituted of any orbital system of electrons however complex'.

I think it may be pedagogically useful to start with the example, integrating the equations in some pedestrian way, for example by numerical computation. The general argument, involving as it does a change of variables, can (I fear) set off premature philosophizing about space and time.

Note that W. Rindler, *Am. J. Phys.* **38**(1970), 1111, finds Larmor insufficiently explicit about time dilation:

'Apparently *no one* before Einstein in 1905 voiced the slightest suspicion that all moving clocks might go slow'.

10

Einstein–Podolsky–Rosen experiments

I have been invited to speak on 'foundations of quantum mechanics' – and to a captive audience of high energy physicists! How can I hope to hold the attention of such serious people with philosophy? I will try to do so by concentrating on an area where some courageous experimenters have recently been putting philosophy to experimental test.

The area in question is that of Einstein, Podolsky, and Rosen[1]. Suppose for example[2,3], that protons of a few MeV energy are incident on a hydrogen target. Occasionally one will scatter, causing a target proton to recoil. Suppose (Fig. 1) that we have counter telescopes T_1 and T_2 which register when suitable protons are going towards distant counters C_1 and C_2. With ideal arrangements registering of both T_1 and T_2 will then imply registering of both C_1 and C_2 after appropriate time decays. Suppose next that C_1 and C_2 are preceded by filters that pass only particles of given

Fig. 1. Proton–proton scattering gedanken experiment.

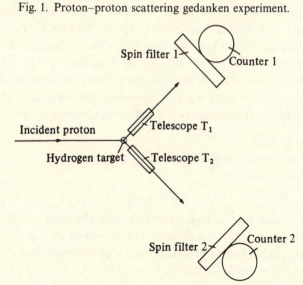

polarization, say those with spin projection $+\frac{1}{2}$ along the z axis. Then one or both of C_1 and C_2 may fail to register. Indeed for protons of suitable energy one and only one of these counters will register on almost every suitable occasion – i.e., those occasions certified as suitable by telescopes[4] T_1 and T_2. This is because proton–proton scattering at large angle and low energy, say a few MeV, goes mainly in S wave. But the antisymmetry of the final wave function then requires the antisymmetric singlet spin state. In this state, when one spin is found 'up' the other is found 'down'. This follows formally from the quantum expectation value

$$\langle \text{singlet} | \sigma_z(1)\sigma_z(2) | \text{singlet} \rangle = -1$$

where $\frac{1}{2}\,\sigma_z(1)$ and $\frac{1}{2}\,\sigma_z(2)$ are the z component spin operators for the two particles.

Suppose now the source–counter distances are such that the proton going towards C_1 arrives there before the other proton arrives at C_2. Someone looking at counter C_1 will not know in advance whether it will or will not register. But once he has noted what happens to C_1 at the appropriate time, he immediately knows what will happen subsequently to C_2, however far away C_2 may be.

Some people find this situation[5] paradoxical. They may, for example, have come to think of quantum mechanics as fundamentally indeterministic. In particular they may have come to think of the result of a spin measurement on an unpolarized particle (and each particle, considered separately, *is* unpolarized here) as utterly indefinite until it has happened. And yet here is a situation where the result of such a measurement is perfectly definitely known in advance. Did it only become determined at the instant when the distant particle passed the distant filter? But how could what happens a long way off change the situation here? Is it not more reasonable to assume that the result was somehow predetermined all along?

I will discuss briefly three ways of responding to this situation, which may be respectively characterized by the following three questions:

> Why worry?
> But is not all this just like classical physics?
> But is it really true?

Why worry?

It can be argued that in trying to see behind the formal predictions of quantum theory we are just making trouble for ourselves. Was not precisely this the lesson that had to be learned before quantum mechanics could be constructed, that it is futile to try to see behind the observed phenomena?

Moreover we learn again from this particular example that we must consider the experimental arrangement as a whole. We must not try to analyse it into separate pieces, with separately localized quotas of indeterminacy. By resisting the impulse to analyze and localize, mental discomfort can be avoided.

This is, as far as I understand it, the orthodox view, as formulated by Bohr[6] in his reply to Einstein, Podolsky, and Rosen. Many people are quite content with it.

But is not all this just as in classical physics?

Similar correlations do indeed exist in classical physics, and surprise nobody. Suppose I take from my pocket a coin and, without looking at it, split it somehow down the middle so that the head and tail are separated. Suppose then that, still without anyone looking, the two different pieces are pocketed by two different people who go on different journeys. The first to look, finding that he has head or tail, will know immediately what the other will subsequently find. Are the quantum mechanical correlations any different? Indeed they are not, according to Einstein[7], if I have understood him correctly. In the example of the coin, the head and the tail were head and tail all along, even while hidden. The person who first looked was just the first to know. But in fact everything was determined from the handing over the pieces (and even before, in fully deterministic classical theory). It is by not explicitly containing the 'hidden variables' reading already head or tail, (or 'up' or 'down'), before observation, that quantum mechanics makes a mystery of a perfectly simple situation. So for Einstein[8]:

> The statistical character of the present theory would then have to be a necessary consequence of the incompleteness of the description of the systems in quantum mechanics, and there would no longer exist any ground for the supposition that a future . . . physics must be based upon statistics . . .

That the apparent indeterminism of quantum phenomena can be simulated deterministically is well known to every experimenter. It is now quite usual, in designing an experiment, to construct a Monte Carlo computer programme to simulate the expected behaviour. The running of the digital computer is quite deterministic – even the so-called 'random' numbers are determined in advance. Every such programme is effectively an *ad hoc* deterministic theory, for a particular set-up, giving the same statistical predictions as quantum mechanics.

It is interesting to follow this up a little in the above case of counter correlations. Let A be a variable which takes the values ± 1 according to whether counter 1 does or does not register. Let $B = \pm 1$ be a similar variable describing the response of counter 2. Let A and B be determined by variables $\lambda, \mu, \nu \ldots$, some of which may be random numbers:

$$A(\lambda, \mu, \nu, \ldots)$$
$$B(\lambda, \mu, \nu, \ldots)$$

There are infinitely many ways of choosing such variables and such functions so that $B = -1$ whenever $A = +1$, and vice versa. The quantum mechanical correlations are then reproduced.

Consider, however, a variation on the experiment. Instead of having both filters pass spins pointing in the z direction, let the two filters be rotated, to pass spins pointing in some other directions. Let the filter associated with the first counter pass spins pointing along some unit vector $\mathbf{a}$, and that associated with the second counter pass spins pointing along some unit vector $\mathbf{b}$. For given values of the hidden variables $\lambda, \mu, \nu, \ldots$ the response A of the first counter may well depend now on the orientation $\mathbf{a}$ of its own filter. But one would not expect A to depend on the orientation $\mathbf{b}$ of the distant second filter. And one could expect the response B of the second counter to depend on the local condition $\mathbf{b}$, but not on the condition $\mathbf{a}$ of the remote instrument:

$$A(\mathbf{a}, \lambda, \mu, \nu, \ldots)$$
$$B(\mathbf{b}, \lambda, \mu, \nu, \ldots)$$

Let the correlation function $P(a, b)$ be defined as the mean value of the product AB:

$$P(a, b) = \overline{A(\mathbf{a}, \lambda, \mu, \nu, \ldots)B(\mathbf{b}, \lambda, \mu, \nu, \ldots)} \tag{1}$$

where the bar denotes averaging over some distribution of variables $\lambda, \mu, \nu, \ldots$

For this more general situation the quantum prediction is

$$P(a, b) = \langle \text{singlet} | \mathbf{a} \cdot \boldsymbol{\sigma}(1) \mathbf{b} \cdot \boldsymbol{\sigma}(2) | \text{singlet} \rangle = -\cos \theta \tag{2}$$

where θ is the angle between $\mathbf{a}$ and $\mathbf{b}$. Can we, by some clever scheme of variables $\lambda, \mu, \nu, \ldots$ and functions A, B, arrange that the average (1) has the value (2)? The answer is 'no'.

Suppose, for example, we arrange that (1) equals (2) for $\mathbf{a} = \mathbf{b}$, i.e., $\theta = 0$:

$$P(a, b) = -1 \quad \text{for} \quad \mathbf{a} = \mathbf{b}$$

Then A and B must have opposite signs every where in the $\lambda, \mu, \nu, \ldots$ space.

Consider now what happens when **a** is varied to some new value **a′**. *B* (which is independent of **a** by hypothesis) does not change for given $\lambda, \mu, \nu, \ldots$ But *A* will change sign at certain points, and these points will contribute $AB = +1$ instead of $AB = -1$ in the average (1). So

$$P(a', a) - P(a, a) = 2\rho$$

where ρ is the total probability of the set of points $\lambda, \mu, \nu, \ldots$ at which *A* changes sign. Now this set of points, at which *A* changes sign when **a** is varied to **a′**, in no way depends on **b**. It follows from (1), and from $B = \pm 1$, that

$$|P(a', b) - P(a, b)| \leqslant 2\rho$$

So of all values **b**, **b** = **a** is that for which *P* varies most rapidly with **a**. Unlike the quantum correlation (2), which is stationary in θ at $\theta = 0$, at the hidden variable correlation (1) must have a *kink* there (Fig. 2).

One could, of course, get the quantum mechanical result from a more general hidden variable representation in which *A* depends on **b** as well as **a**, or *B* on **a** as well as **b**:

$$A(\mathbf{a}, \mathbf{b}, \lambda, \mu, \nu, \ldots)$$
$$B(\mathbf{a}, \mathbf{b}, \lambda, \mu, \nu, \ldots)$$

But this would make the behaviour of a counter dependent on what is done at a distant place. This would seen strange enough with **a** and **b** constant, but suppose now that these settings vary with time. Then according to quantum mechanics the relevant values of **a** and **b** are those obtained when

Fig. 2. Behaviour of correlation *P* near $\theta = 0$, $P = -1$.

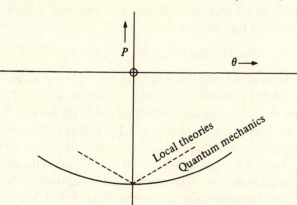

the particles pass through the corresponding filters. Suppose for example we arrange that the two passages are simultaneous. Then A (or B) would have to depend *instantaneously* on the setting $\mathbf{b}$ (or $\mathbf{a}$) of the distant instrument. The causal dependence would have to propagate faster than light.

So all this is not at all just like classical physics. Einstein argued that the EPR correlations could be made intelligible only by completing the quantum mechanical account in a classical way. But detailed analysis shows that any classical account of these correlations has to contain just such a 'spooky action at a distance'[9] as Einstein could not believe in:

> But on one supposition we should, in my opinion, absolutely hold fast: the real factual situation of the system S_2 is independent of what is done with the system S_1 which is spatially separated from the former[10].

If nature follows quantum mechanics in these correlations, then Einstein's conception of the world is untenable.

But is it really true?

Well, *does* nature follow quantum mechanics in these matters? It might be argued that the very general and very remarkable success of quantum mechanics makes it pointless to do special experiments on these correlations. We will just find, after a lot of trouble, that quantum mechanics is again right. But it can also be argued that the great success of quantum mechanics, in so far as it differs from classical mechanics, is on the microscopic scale. Here, on the other hand, we are concerned with specifically quantum phenomena on the macroscopic scale.

The present movement to check these things experimentally started with the key paper of Clauser, Holt, Horne, and Shimony[11]. From the basic representation (1) they showed that

$$|P(a,b) - P(a,b')| + |P(a',b) + P(a',b')| \leqslant 2 \tag{3}$$

Here P is the counting correlation already defined, a and a' are alternative settings of the first polarizer, and b and b' alternative settings of the second. It is readily seen that the quantum mechanical P, (2), for well chosen a, a' b, b', violates (3) by a factor as large as $\sqrt{2}$. It is in terms of this very practical 'locality inequality' that the various experiments have been interpreted.

Unfortunately it is not at present possible to approach the conditions of the ideal critical experiment. Real counters, real polarization analyzers, and real geometrical arrangements, are together so inefficient that the quantum mechanical correlations are greatly diluted. The counters seldom say 'yes, yes', usually say 'no, no', and say 'yes no' with a frequency only weakly dependent on the polarizer settings. In these conditions

$$P(a,b) = 1 - (\delta(a,b))^2$$

where δ is small and weakly dependent on the arguments a, b. The inequality (3) is then trivially satisfied. So it is only by allowing (in effect) for various inefficiencies in conventional ways, and so *extrapolating* from the real results to hypothetical ideal results, that the various experiments can be said to 'test' the inequality. But the results are nevertheless of great interest. Compensating failures could be imagined, of the conventional quantum mechanics of spin correlations and of the conventional phenomenology of the instruments, which would make the practical experiments irrelevant. But that would seem an extraordinary conspiracy.

Of these experiments only one is concerned with the low energy pp scattering of the above gedanken experiment. It is that of Lamehi-Rachti and Mittig at Saclay[12]. Protons of 14 MeV lab energy are scattered at a lab angle of 45°, and spin correlation of scattered and recoil protons measured. They do not have the ideal yes–no polarization filters of the gedanken experiment. Instead they analyze polarization by secondary scattering on Carbon. Nor do they have the telescopes T_1 and T_2 to tell when there are indeed suitable particles going towards the counters. This also lengthens the extrapolation from real to ideal experiment. Nevertheless if there were some tendency for the singlet spin state to dissipate somehow with macroscopic separation of the particles, it should show up, barring conspiracy, in such an experiment. The preliminary results show no such effect. They agree with quantum mechanics and disagree (in the sense of a certain extrapolation) with the locality inequality.

All the other experiments have been done with pairs of photons rather than spin half particles. In the theory the two linear polarization states of each photon replace the two spin states of each spin $\frac{1}{2}$ particle. Suitably correlated photon pairs arise in the annihilation of slow positrons with electrons. Again there are no very efficient polarization filters. The experimenters have to resort to Compton scattering of the photons; according to quantum mechanics the polarization correlations are then translated into angular correlations. Such experiments have been done at Columbia[13] (Kasday, Ullman, and Wu) and at Catania[14] (Faraci, Gutkowski, Notarrigo, and Pennisi). The Columbia result is in agreement with quantum mechanics, and (in the extrapolated sense) in significant disagreement with the inequality. The reverse is the case for the Catania experiment.

The reasons for this discrepancy between the two experiments are not known, as far as I can tell.

For optical photons, in contrast with the energetic photons of positron annihilation, efficient polarization filters *are* available – namely birefringent crystals and 'piles-of-plates'. Moreover suitably correlated photon pairs are produced in certain atomic cascades. Consider for example a two photon cascade in which initial and final atomic states have zero angular momentum. When the two photons come off back to back their helicities must be so correlated that there is no net angular momentum about their common direction of motion. There is a corresponding correlation of linear polarization states. Unfortunately the photons do not always come off back to back, for the residual atom can take up momentum. Very often then a 'no' from a counter has no significance for polarization, but just means that no photon has gone that way. This problem could be eliminated in principle by suitable telescopes T to veto the uninteresting cases. But this has not been possible in practice. The significance of 'no' from a counter is further diminished in these experiments by the very low efficiencies of the photon counters. So there is no question of actually realizing a system which violates the locality inequality. But such experiments do test whether the quantum polarization correlations persist over macroscopic distances. Experiments have been done by Clauser and Freedman[15], on a cascade in Calcium, by Holt and Pipkin[16] and by Clauser[17] on a cascade in Mercury, and by Fry[18] on another cascade in Mercury. Three of these four experiments confirm quantum mechanics very nicely and (in the sense of some extrapolation) disagree significantly with the locality inequality. But for Holt and Pipkin the reverse is true. It is not understood why this experiment disagrees with the very similar one of Clauser.

Now these experiments do not test at all what was said to be the most striking feature of the quantum correlations. This was their dependence only on the instantaneous settings, during the passage of the particles, of the polarization filters. It is therefore of very great interest that an atomic cascade experiment is now under way in which *the settings of the polarizers are changed while the photons are in flight*. Clauser[19] suggested that this might be done by the use of something like Kerr cells. But according to Aspect[20] such cells heat up too quickly and are of too low transmission to be useful in practice. His idea is to replace each filter–counter combination by a pair of such combinations with differently oriented filters. He thinks that he can bring one or other orientation into play by a switching device that can rapidly redirect the incident photon from one filter to the other. He believes that such switching can be effected by the generation of ultrasonic standing waves on which the photon undergoes Bragg reflection. If this experiment gives the expected result it will be a confirmation of what is, to my mind, in the light of the locality analysis[21], one of the most extraordinary predictions of quantum theory.

I think that future generations should be grateful to those who bring these matters out of the realm of gedanken experiment into that of real experiment. Moreover several of the real experiments are of great elegance. To hear of them (not in schematic terms from a theorist but in real terms from their authors) is, to borrow a phrase from Professor Gilberto Bernardini, a spiritual experience.

Appendix: Einstein and hidden variables

I had for long thought it quite conventional and uncontroversial to regard Einstein as a proponent of hidden variables, and indeed[22] as 'the most profound advocate of hidden variables'. And so I had on several occasions appealed to the authority of Einstein to legitimise an interest in this question. But in so doing I have been accused, by Max Jammer[5] in his very valuable book: *The Philosophy of Quantum Mechanics*, of misleading the public:

> One of the sources of erroneously listing Einstein among the proponents of hidden variables was probably J. S. Bell's widely read paper: On the Einstein–Podolsky–Rosen Paradox, *Physics* **1**, 195–200 (1964), which opened with the statement: 'The paradox...was advanced as an argument that quantum mechanics...should be supplemented by additional variables.'...Einstein's remarks in his 'Reply to Criticisms' (Ref. 4–9, p. 672), quoted by Bell in support of his thesis, are certainly no confession of the belief in the necessity of hidden variables.

The remark of Einstein which I had quoted was this:

> But on one supposition we should, in my opinion, absolutely hold fast: the real factual situation of the system S_2 is independent of what is done with the system S_1, which is spatially separated from the former.

The object of this quotation was to recall Einstein's deep commitment to realism and locality, the axioms of the EPR paper. And the quotation was not from p. 672 of Einstein's 'Reply to Criticisms', but from p. 85 of his 'Autobiographical Notes' in the same volume[23]. But turning to p. 672, I find the following:

> Assuming the success of efforts to accomplish a complete physical description, the statistical quantum theory would, within the framework of future physics, take an approximately analogous position to the statistical mechanics within the framework of classical mechanics. I am rather firmly convinced that the development of theoretical physics will be of this type; but the path will be lengthy and difficult.

This seems to me a rather clear commitment to what is usually meant by hidden variables[24].

Other similarly clear statements are readily found[25]:

> I am, in fact, firmly convinced that the essentially statistical character of contemporary quantum theory is solely to be ascribed to the fact that this (theory) operates with an incomplete description of physical systems.

Moreover, the Einstein–Podolsky–Rosen paper *did* have the title: 'Can Quantum Mechanical Description of Physical Reality be Considered Complete?' And it did end with:

> While we have thus shown that the wave function does not provide a complete description of the physical reality, we left open the question of whether or not such a description exists. We believe, however, that such a theory is possible.

It seems to me then beyond dispute that there was at least one Einstein, that of the EPR paper and the Schilpp volume, who was fully committed to the view that quantum mechanics was incomplete and should be completed – which is the hidden variable programme. Max Jammer seems not to have found this Einstein, but claims to have found another. As evidence he cites phrases from private letters, an oral tradition, and Einstein's well-known commitment to classical field theory.

Now the belief in classical field theory, in 'Continuous functions in the four dimensional (continuum) as basic concepts of the theory[26]', in no way excludes belief in 'hidden' variables. It can be seen rather as a particular conception of those variables.

The oral tradition was that Einstein expected quantum mechanics ultimately to come in conflict with experiment. But if such an expectation were to exclude him from the list of proponents of hidden variables, I doubt it anyone could be left on it. If such a list were compiled I think it would be of people concerned to reproduce the experimentally confirmed aspects of quantum mechanics but eager to find in their investigations some hint as to where a critical experiment might be sought. Indeed few would expect the ultimate vindication of quantum mechanics (on the statistical level) so strongly as Einstein himself on one occasion[27]: 'The formal relations which are given in this theory – i.e., its entire mathematical formalism – will probably have to be contained, in the form of logical inferences, in every useful future theory'.

The quotations from private letters are of negative reactions by Einstein to the very particular 1952 hidden variables of Bohm. This scheme reproduced completely, and rather trivially, the whole of nonrelativistic quantum mechanics. It had great value in illuminating certain features of the theory, and in putting in perspective various 'proofs' of the impossibility of a hidden variable interpretation. But Bohm himself did not think of it as in any way final. Jammer could have added to his quotations the following, from a letter from Einstein to Born[6]:

> Have you noticed that Bohm believes (as de Broglie did, by the way, 25 years ago) that he is able to interpret the quantum theory in deterministic terms? That way seems too cheap to me.

On which Born comments:

> Although this theory was quite in line with his own ideas,...

So Born also had listed Einstein as a proponent of hidden variables. I think he was right.

Notes and references

1 A. Einstein, B. Podolsky and N. Rosen, *Phys. Rev.* **47**, 777 (1935).
2 D. Bohm, *Quantum Theory*, Englewood Cliffe, N.J. (1951).
3 A. Peres and P. Singer, *Nuovo Cimento* **15**, 907 (1960); R. Fox, *Lettere al Nuovo Cimento* **2**, 656 (1971).
4 It is assumed that these telescopes do not affect proton spin.
5 M. Jammer, *The Philosophy of Quantum Mechanics*, Wiley, N.Y. (1974). Chapters 6 and 7 give a comprehensive account of the history (and prehistory) of the EPR paradox.
6 N. Bohr, Discussions with Einstein, in Ref. 23.
7 Appendix.
8 A. Einstein, in Ref. 23, p. 87.
9 A. Einstein, in Ref. 28, p. 158.
10 A. Einstein, in Ref. 23, p. 85.
11 J. F. Clauser, R. A. Holt, M. A. Horne and A. Shimony, *Phys. Rev. Lett.* **23**, 880 (1969).
12 M. Lamehi-Rachti and W. Mittig, *Phys. Rev.* **D14**, 2543 (1976).
13 L. R. Kasday, J. D. Ullman and C. S. Wu, *Nuovo Cimento* **25B**, 633 (1975).

14 G. Faraci, D. Gutkowski, S. Notarrigo and A. R. Pennisi, *Lettere al Nuovo Cimento* **9**, 607 (1974).
15 J. F. Clauser and S. J. Freedman, *Phys. Rev. Lett.* **28**, 938 (1972).
16 F. M. Pipkin, *Adv. Atomic and Mol. Phys.* **14**, 281 (1978).
17 J. F. Clauser, *Phys. Rev. Lett.* **36**, 1223 (1976).
18 E. S. Fry and R. C. Thompson, *Phys. Rev. Lett.* **37**, 465 (1976).
19 As reported by A. Shimony, Ref. 22.
20 A. Aspect, *Phys. Lett.* **A54**, 117 (1975), *Phys. Rev.* **D14**, 1944 (1976).
21 For simplicity, in this paper we followed up the consequences of determinism, which is required by locality only in the case of ideal perfect correlations. But (3) holds in a much wider class of theories, local but indeterministic. See, for example, and references therein: J. F. Clauser and M. A. Horne, *Phys. Rev.* **D10**, 526 (1974); B. D'Espagnat, *Phys. Rev.* **D11**, 1424 (1975); and *Conceptual Foundations of Quantum Mechanics,* Benjamin, new edition (1976); J. S. Bell, *The Theory of Local Beables,* CERN, TH 2053 (1975), in GIFT (1975) Proceedings and Epistemological Letters March 1976.
22 A. Shimony, in *Foundations of Quantum Mechanics,* B. D'Espagnat, Ed. Academic Press, N.Y., London (1971), p. 192, quoted with disapproval by M. Jammer ref. 5.
23 P. A. Schilpp, Ed., *Albert Einstein, Philosopher-Scientist,* Tudor, N.Y. (1949).
24 The usual nomenclature, *hidden* variables, is most unfortunate. Pragmatically minded people can well ask *why bother about hidden entities that have no effect on anything?.* Of course, every time a scintillation occurs on screen, every time an observation yields one thing rather than another, the value of a *hidden* variable is revealed. Perhaps *uncontrolled* variable would have been better, for these variables, by hypothesis, for the time being, cannot be manipulated at will by us.
25 Ref. 23, p. 666. See also Einstein's introductory remarks in Louis de Broglie, *Physicien et Penseur,* Albin Michel, Paris (1953), p. 5, and letters 81, 84, 86, 88, 97, 99, 103, 106, 108, 110, 115 and 116, in Ref. 28.
26 Ref. 23, p. 675.
27 Ref. 23, p. 667.
28 M. Born, Ed., *The Born-Einstein Letters,* p. 192, Macmillan, London (1971).

11

The measurement theory of Everett and de Broglie's pilot wave

In 1957 H. Everett published a paper setting out what seemed to be a radically new interpretation of quantum mechanics[1]. His approach has recently received increasing attention[2]. He did not refer to the ideas of de Broglie of thirty years before[3] nor to the intervening elaboration of those ideas by Bohm[4]. Yet it will be argued here that the elimination of arbitrary and inessential elements from Everett's theory leads back to, and throws new light on, the concepts of de Broglie[5].

Everett was motivated by the notion of a quantum theory of gravitation and cosmology. In a thoroughly quantum cosmology, a quantum mechanics of the whole world, the wave function of the world could not be interpreted in the usual way. For this usual interpretation refers only to the statistics of measurement results for an observer intervening from outside the quantum system. When that system is the whole world, there is nothing outside. This situation presents no particular difficulty for the traditional (or 'Copenhagen') philosophy, which holds that a classical conception of the macroscopic world is logically prior to the quantum conception of the microscopic. The microscopic world is described by wave functions which are determined by and have implications for macroscopic phenomena in experimental set-ups. These macroscopic phenomena are described in a perfectly classical way (in the language of '*be*-ables'[6] rather than 'observables', so that there is no question of an endless chain of observers observing observers observing....). There is of course no sharply defined boundary between what is to be treated as microscopic and what as macroscopic, and this introduces a basic vagueness into fundamental physical theory. But this vagueness, because of the immense difference of scale between the atomic level where quantum concepts are essential and the macroscopic level where classical concepts are adequate, is quantitatively insignificant in any situation hitherto envisaged. So, it is quite acceptable to many people. It is not surprising then that such a consistent traditionalist as L. Rosenfeld has gone so far as to suggest[7] that a quantum theory of gravitation may be unnecessary. The only gravitational phenomena we actually *know* are of macroscopic scale and involve very many atoms. So we only *need* the concept of gravitation on this classical level, whose separate logical status is anyway fundamental in the traditional view. Nevertheless, I think that most contemporary physicists would regard any purely classical theory of gravitation as provisional, and hold that any really adequate theory must be applicable, in principle, also on the microscopic level – even if its effects there are negligibly small[8]. Many of these same contemporary physicists are perfectly complacent about the vague division of the world into classical macroscopic and quantum microscopic inherent in contemporary (i.e., traditional) quantum theory. This mixture of concern on the one hand and complacency on the other is in my opinion less admirable than the clear headed and systematic complacency of Rosenfeld.

Everett was complacent neither about gravitation nor quantum theory. As a preliminary to a synthesis of the two he sought to interpret the notion of a wave function for the world. This world certainly contains instruments that can detect, and record macroscopically, microscopic and other phenomena. Let A be the recording part, or 'memory', of such a device, or of a collection of such devices, and let B be the rest of the world. Let the co-ordinates of A be denoted by a, and of B by b. Let $\phi_n(a)$ be a complete set of states for A. Then, one can expand the world wave function $\psi(a, b, t)$ at some time t in terms of the ϕ_n:

$$\psi(a, b, t) = \sum_n \phi_n(a)\chi_n(b, t) \tag{E}$$

We will refer to the norm of χ_n

$$\int db\, |\chi_n(b, t)|^2$$

as the 'weight' of ϕ_n in the expansion. As an example A might be a photographic plate that can record the passage of an ionizing particle in a pattern of blackened spots. The different patterns of blackening correspond to different states ϕ_n. Then it can be shown[9] along lines laid down long ago by Mott and Heisenberg, that the only states ϕ_n with appreciable weight are those in which the blackened spots form essentially a linear sequence, in which the blackening of neighbouring plates, or of different parts of the same plate, are consistent with one another, and so on. In the same way Everett, allowing A to be a more complicated memory, such as that of a computer (or even a human being), or a collection of such memories, shows that only those states ϕ_n have appreciable weight in which the memories agree on a more or less coherent story of the kind we have experience of. All this is neither new nor controversial. The novelty is in the emphasis on memory contents as the essential material of physics and in the interpretation which Everett proceeds to impose on the expansion E.

An exponent of the traditional view, if he allowed himself to contemplate a wave function of the world, would probably say the following. Once a macroscopic record has been formed we are concerned with fact rather than possibility, and the wave function must be adjusted to take account of this. So from time to time the wave function is 'reduced'

$$\psi \to N \sum{}' \phi_n(a)\chi_n(b,t) \qquad (E')$$

where (N being a renormalization factor) the restricted summation $\sum{}'$ is over a group of states ϕ_n which are 'macroscopically indistinguishable'. The complete set of states is divided into many such groups, and the reduction to a particular group occurs with probability proportional to its total weight

$$\sum{}' \int db \, |\chi_n|^2.$$

He will not be able to say just when or how often this reduction should be made, but would be able to show by analyzing examples that the ambiguity is quantitatively unimportant in practice. Everett disposes of this vaguely defined suspension of the linear Schrödinger equation with the following bold proposal: it is just an illusion that the physical world makes a particular choice among the many macroscopic possibilities contained in the expansion; they are *all* realized, and no reduction of the wave function occurs. He seems to envisage the world as a multiplicity of 'branch' worlds, one corresponding to each term $\phi_n\chi_n$ in the expansion. Each observer has representatives in many branches, but the representative in any particular branch is aware only of the corresponding particular memory state ϕ_n. So he will remember a more or less continuous sequence of past 'events', just as if he were living in a more or less well defined single branch world, and have no awareness of other branches. Everett actually goes further than this, and tries to associate each particular branch at the present time with some particular branch at any past time in a tree-like structure, in such a way that each representative of an observer has actually lived through the particular past that he remembers. In my opinion this attempt does not succeed[9] and is in any case against the spirit of Everett's emphasis on memory contents as the important thing. We have no access to the past, but only to present memories. A present memory of a correct experiment having been performed should be associated with a present memory of a correct result having been obtained. If physical theory can account for such correlations in present memories it has done enough – at least in the sprit of Everett.

Rejecting the impulse to dismiss Everett's multiple universe as science fiction, we raise here a couple of questions about it.

The first is based on this observation: there are infinitely many different expansions of type E, corresponding to the infinitely many complete sets ϕ_n. Is there then an additional multiplicity of universes corresponding to the infinitely many ways of expanding, as well as that corresponding to the infinitely many terms in each expansion? I think (I am not sure) that the answer is no, and that Everett confines his interpretation to a particular expansion. To see why suppose for a moment that A is just an instrument with two readings 1 and 2, the corresponding states being ϕ_1 and ϕ_2. Instead of expanding in ϕ_1 and ϕ_2 we could, as a mathematical possibility, instead expand in

$$\phi_{\pm} = (\phi_1 \pm \phi_2)/\sqrt{2} \quad \text{or} \quad \phi'_{\pm} = (\phi_1 \pm i\phi_2)/\sqrt{2}.$$

In each of these states the instrument reading takes no definite value, and I do not think Everett wishes to have branches of this kind in his universe. To formalize his preference let us introduce an instrument reading operator R:

$$R\phi_n = n\phi_n$$

and operators Q and P similarly related to $\phi_{\pm}$ and $\phi'_{\pm}$. Then we can say that Everett's structure is based on an expansion in which instrument readings R, rather than operators like Q or P, are diagonalized. This preference for a particular set of operators is not dictated by the mathematical structure of the wave function ψ. It is just added (only tacitly by Everett, and only if I have not misunderstood) to make the model reflect human experience. The existence of such a preferred set of variables is one of the elements in the close correspondence between Everett's theory and de Broglie's – where the positions of particles have a particular role.

The second question grows out of the first: if instrument readings are to be given such a fundamental role should we not be told more exactly what an instrument reading is, or indeed, an instrument, or a storage unit in a memory, or whatever? In dividing the world into pieces A and B Everett is indeed following an old convention of abstract quantum measurement theory, that the world does fall neatly into such pieces – instruments and systems. In my opinion this is an unfortunate convention. The real world is made of electrons and protons and so on, and as a result the boundaries of natural objects are fuzzy, and some particles in the boundary can only doubtfully be assigned to either object or environment. I think that fundamental physical theory should be so formulated that such artificial divisions are manifestly inessential. In my opinion Everett has not given such a formulation – and de Broglie has.

So we come finally to de Broglie. Long ago he faced the basic duality of quantum theory. For a single particle the mathematical wave extends over space, but the experience is particulate, like a scintillation on a screen. For a complex system, ψ extends over the whole configuration space, and over all n in expansions like (E), but experience has a particular character, like the reduced expansion (E'). De Broglie made the simple and natural suggestion: the wave function ψ is not a complete description of reality, but must be supplemented by other variables. For a single particle he adds to the wave function $\psi(\mathbf{r}, t)$ a particle coordinate $\mathbf{x}(t)$ – the instantaneous position of the localized particle in the extended wave. It changes with time according to

$$\dot{\mathbf{x}} = \left[\operatorname{Im} \psi^*(\mathbf{x}, t) \frac{\partial}{\partial \mathbf{x}} \psi(\mathbf{x}, t) \right] \bigg/ |\psi(\mathbf{x}, t)|^2. \qquad (G)$$

In an ensemble of similar situations $\mathbf{x}$ is distributed with weight $|\psi(\mathbf{x}, t)|^2 \, d\mathbf{x}$, a situation which follows from (G) for all t if it holds at some t. To make a model of the world, a simple world consisting just of many non-relativistic particles, we have only to extend these prescriptions from 3 to $3N$ dimensions, where N is the total number of particles. In this world the many-body wave function obeys exactly a many-body Schrödinger equation. There is no 'wave function reduction', and all terms in expansions like E are retained indefinitely. Nevertheless the world has a definite configuration $(\mathbf{x}_1, \mathbf{x}_2, \mathbf{x}_3 \ldots)$ at every instant, changing according to the $3N$ dimensional version of (G). This model is like Everett's in employing a world wave function and an exact Schrödinger equation, and in superposing on this wave function an additional structure involving a preferred set of variables. The main differences seem to me to be these.

(1) Whereas Everett's special variables are the vaguely anthropocentric instrument readings, de Broglie's are related to an assumed microscopic structure of the world. The macroscopic features of direct interest to human beings, like instrument readings, can be brought out by suitably coarse-grained averaging, but the ambiguities in doing so do not enter the fundamental formulation.

(2) Whereas Everett assumes that *all* configurations of his special variables are realized at any time, each in the appropriate branch universe, the de Broglie world has a *particular* configuration. I do not myself see that anything useful is achieved by the assumed existence of the other branches of which I am not aware. But let he who finds this assumption inspiring

make it; he will no doubt be able to do it just as well in terms of the xs as in terms of the Rs.

(3) Whereas Everett makes no attempt, or only a half-hearted one, to link successive configurations of the world into continuous trajectories, de Broglie does just this in a perfectly deterministic way (G). Now these trajectories of de Broglie, innocent as (G) may look in the configuration space, are really very peculiar as regards locality in ordinary three-space[9]. But we learn from Everett that if we do not like these trajectories we can simply leave them out. We could just as well redistribute the configuration $(\mathbf{x}_1, \mathbf{x}_2, \ldots)$ at random (with weight $|\psi|^2$) from one instant to the next. For we have no access to the past, but only to memories, and these memories are just part of the instantaneous configuration of the world.

Does this final synthesis, omitting de Broglie's trajectories and Everett's other branches, make a satisfactory formulation of fundamental physical theory? Or rather would some variation of it based on a relativistic field theory? It is logically coherent, and does not need to supplement mathematical equations with vague recipes. But I do not like it. Emotionally, I would like to take more seriously the past of the world (and of myself) than this theory would permit. More professionally, I am uneasy about the possibility of incorporating relativity in a profound way. No doubt it would be possible to ensure memory of a null result for the Michelson–Morley experiment and so on. But could the basic reality be other than the state of world, or at least a memory, extended in space at a single time – defining a preferred Lorentz frame? To try to elaborate on this would only be to try to share my confusion.

Notes and references

1 Everett, H., *Revs. Modern Phys.* **29**, 454 (1957); see also Wheeler, J. A., *Revs. Modern Phys.* **29**, 463 (1957).

2 See for example:
de Witt, B. S. and others in *Physics Today* **23** (1970), No. 9, 30 and **24**, No. 4, 36 (1971) and references therein. Ideas like those of Everett have also been set out by Cooper, L. N. and van Vechten, D., *American J. Phys.* **37**, 1212 (1969) and by L. N. Cooper in his contribution to the Trieste symposium in honour of P. A. M. Dirac, September 1972.

3 For a systematic exposition see: de Broglie, L., *Tentative d'Interprétation Causale et Non-linéaire de la Mécanique Ondulatoire*, Gauthier-Villars, Paris, (1956).

4 Bohm, D., *Phys. Rev.* **85**, 166, 180, (1952).

5 This thesis has already been presented in my contribution to the international colloquium on issues in contemporary physics and philosophy of science, Penn. State University, September 1971, CERN TH. 1424. That paper is referred to for more details of several arguments, but the opportunity has been taken here to expand on some points only mentioned there.

6 Bell, J. S., contribution to the Trieste symposium in honour of P. A. M. Dirac, CERN TH. 1582, September 1972[10].

7 Rosenfeld, L., *Nuclear Phys.* **40**, 353 (1963).
   G. F. Chew has suggested that the *electromagnetic* interaction must be considered apart (although not of course left unquantized) because of its macroscopic role in observation. (*High Energy Physics*, Les Houches, 1965, ed. by C. de Witt and M. Jacob. Gordon and Breach (1965)).

8 It is beside the present point that microscopic gravitation might not in fact be quantitatively unimportant; see, for example, the contribution of A. Salam to the Trieste symposium in honour of P. A. M. Dirac, September 1972[10].

9 For details see the paper referred to in note 5.

10 *The Physicist's Conception of Nature*, Ed. by J. Mehra, Dordrecht, Reidel (1973).

12

Free variables and local causality

It has been argued[1] that quantum mechanics is not locally causal and cannot be embedded in a locally causal theory. That conclusion depends on treating certain experimental parameters, typically the orientations of polarization filters, as free variables. Roughly speaking it is supposed that an experimenter is quite free to choose among the various possibilities offered by his equipment. But it might be that this apparent freedom is illusory. Perhaps experimental parameters and experimental results are both consequences, or partially so, of some common hidden mechanism. Then the apparent non-locality could be simulated.

This possibility is the starting point of a paper by Clauser, Horne and Shimony[2] (CHS hereafter), which is valuable in particular for a careful mathematical formulation of the assumption which excludes such a conspiracy. In this connection they severely criticize my own 'theory of local beables'[1] (B hereafter). Much of their criticism is perfectly just. In B there were jumps[3] in the argument, and the assumption in question was not stated at the appropriate place, but only later and inadequately. However, I do not agree with CHS that this assumption, when carefully formulated, is an unreasonable one.

I will organize these remarks around the three phrases in which I belatedly formulated the hypothesis in B, Section 8.

1 '*It has been assumed that the settings of instruments are in some sense free variables …*'

For me this means that the values of such variables have implications only in their future light cones. They are in no sense a record of, and do not give information about, what has gone before. In particular they have no implications for the hidden variables v in the overlap of the backward light cones:

$$\{v|a,b,c\} = \{v|a',b,c\} = \{v|a,b',c\} = \{v|a',b',c\} \tag{1}$$

This, as explained by CHS, is what is used in the mathematical analysis.

The bracket symbol denotes the probability of particular values v given particular values a, b, c where c lists non-hidden variables in the overlap of the backward light cones of two instruments, and a and b list non-hidden variables in the remainders of those light cones. The lists a and a' are supposed to differ in the setting of the first instrument, while b and b' are supposed to differ in the setting of the second instrument.

Note that instead of (1) CHS write, probably interpreting the symbols a little differently

$$\{v|a, b, c\} = \{v|c\}$$

With my notation, where a and b are lengthy lists of variables describing the situation outside the overlap, this would be much stronger than (1) – and not reasonable at all.

2 '... *say at the whim of experimenters* ...'

Here I would entertain the hypothesis that experimenters have free will. But according to CHS it would not be permissible for me to justify the assumption of free variables 'by relying on a metaphysics which has not been proved and which may well be false'. Disgrace indeed, to be caught in a metaphysical position! But it seems to me that in this matter I am just pursuing my profession of theoretical physics.

I would insist here on the distinction between analyzing various physical theories, on the one hand, and philosophising about the unique real world on the other hand. In this matter of causality it is a great inconvenience that the real world is given to us once only. We cannot know what would have happened if something had been different. We cannot repeat an experiment changing just one variable; the hands of the clock will have moved, and the moons of Jupiter. Physical theories are more amenable in this respect. We can *calculate* the consequences of changing free elements in a theory, be they only initial conditions, and so can explore the causal structure of the theory. I insist that B is primarily an analysis of certain kinds of physical theory.

A respectable class of theories, including contemporary quantum theory as it is practised, have 'free' 'external' variables in addition to those internal to and conditioned by the theory. These variables are typically external fields or sources. They are invoked to represent experimental conditions. They also provide a point of leverage for 'free willed experimenters', if reference to such hypothetical metaphysical entities is permitted. I am inclined to pay particular attention to theories of this kind, which seem to me most simply related to our everyday way of looking at the world.

Of course there is an infamous ambiguity here, about just what and where the free elements are. The fields of Stern–Gerlach magnets could be treated as external. Or such fields and magnets could be included in the quantum mechanical system, with external agents acting only on external knobs and switches. Or the external agents could be located in the brain of the experimenter. In the latter case the setting of the instrument is *not* itself a free variable. It is only more or less closely correlated with one, depending on how accurately the experimenter effects his intention. As he puts out his hand to the knob, his hand may shake, and may shake in a way influenced by the variables *v*. Remember, however, that the disagreement between locality and quantum mechanics is large – up to a factor of $\sqrt{2}$ in a certain sense. So some hand trembling can be tolerated without much change in the conclusion. Quantification of this would require careful epsilonics.

3 '... or at least not determined in the overlap of the backward light cones'

Here I must concede at once that the hypothesis becomes quite inadequate when weakened in this way. The theorem no longer follows. I was mistaken.

At this point I had in mind the possibility of exploiting the freedom, in conventional physical theories, of initial conditions. I am now embarrassed not only by the inadequacy of this particular phrase in the hypothesis, but also by the necessity of paying attention in such a study to the creation of the world[4].

Let me instead then weaken the hypothesis in a different and more practical way.

4 '... or at least effectively free for the purpose at hand'

Suppose that the instruments are set at the whim, not of experimental physicists, but of mechanical random number generators. Indeed it seems less impractical to envisage experiments of this kind[5], with space-like separation between the outputs of two such devices, than to hope to realize such a situation with human operators. Could the outputs of such mechanical devices reasonably be regarded as sufficiently free for the purpose at hand? I think so.

Consider the extreme case of a 'random' generator which is in fact perfectly deterministic in nature – and, for simplicity, perfectly isolated. In such a device the complete final state perfectly determines the complete initial state – nothing is forgotten. And yet for many purposes, such a device is precisely a 'forgetting machine'. A particular output is the result of combining so many factors, of such a lengthy and complicated dynamical chain, that it is quite extraordinarily sensitive to minute variations of any one of many initial conditions. It is the familiar paradox of classical statistical mechanics that such exquisite sensitivity to initial conditions is practically equivalent to complete forgetfulness of them. To illustrate the point, suppose that the choice between two possible outputs, corresponding to a and a', depended on the oddness or evenness of the digit in the millionth decimal place of some input variable. Then fixing a or a' indeed fixes something about the input – i.e., whether the millionth digit is odd or even. But this peculiar piece of information is unlikely to be the vital piece for any distinctively different purpose, i.e., it is otherwise rather useless. With a physical shuffling machine, we are unable to perform the analysis to the point of saying just what peculiar feature of the input is remembered in the output. But we can quite reasonably assume that it is not relevant for other purposes. In this sense the output of such a device is indeed a sufficiently free variable for the purpose at hand. For this purpose the assumption (1) is then true enough, and the theorem follows.

Arguments of this kind are advanced by CHS in defending the corresponding assumption in the Clauser–Horne analysis. I do not know why they should be considered less relevant here.

Of course it might be that these reasonable ideas about physical randomizers are just wrong – for the purpose at hand. A theory may appear in which such conspiracies inevitably occur, and these conspiracies may then seem more digestible than the non-localities of other theories. When that theory is announced I will not refuse to listen, either on methodological or other grounds. But I will not myself try to make such a theory.

Notes and references

1 J. S. Bell, *Epistemological Letters*, March 1976, and references therein. Reprinted in *Dialectica* **39** (1985).
2 A. Shimony, M. A. Horne and J. F. Clauser, *Epistemological Letters*, October 1976. Reprinted in *Dialectica* **39** (1985).
3 In particular CHS complain of a difficulty in connection with (14) and (15) in B. What is missing there is the remark that on the right-hand sides the averaging is over μ and μ', and λ and λ', separately. The operation is, explicitly,

$$\int d\lambda\, d\lambda'\, d\mu\, d\mu'\, d\nu \{\lambda|a,c,\nu\} \{\lambda'|a',c,\nu\} \{\mu|b,c,\nu\} \{\mu'|b',c,\nu\} \{\nu|a,b,c,\}$$

According to Eq. (1) above a and/or b in the last factor may be replaced by a' and/or b' respectively. As applied for example to

$$p((\lambda,a),(\mu,b),(\nu,c))$$

which does not depend on λ' or μ', two integrations are trivial, leaving

$$\int d\lambda\, d\mu\, dv\{\lambda|a,c,v\}\,\{\mu|b,c,v\}\,\{v|a,b,c,\}$$

$$\equiv \int d\lambda\, d\mu\, dv\{\lambda|a,b,\mu,c,v\}\,\{\mu|a,b,c,v\}\,\{v|a,b,c\}$$

(using locality)

$$\equiv \int d\lambda\, d\mu\, dv\{\lambda,\mu,v|a,b,c\}$$

which is the averaging involved in defining $P(a,b,c)$. However, I agree with CHS that an earlier style[6], averaging over λ and μ before forming the inequality, is simpler.

4 The invocation in Ref. 1 of a *complete* account of the overlap of backward light cones is embarrassing in a related way, whether going back indefinitely or to a finite creation time – which might, by the way, have even been a creation *point*, with all backward light cones confused. R. P. Feynman in particular objected to the concept of a complete history being involved. In a more careful discussion the notion of completeness should perhaps be replaced by that of sufficient completeness for a certain accuracy, with suitable epsilonics.

5 Important progress in this direction is being made by A. Aspect, *Physical Review* **D14**, 1944 (1976).

6 J. S. Bell, in *Proceedings of the International School of Physics*, Enrico Fermi, Course IL, Varenna 1970.

13

Atomic-cascade photons and quantum-mechanical nonlocality

It has been feared that television is responsible for the disturbing decline of birth rate in France. It is quite unclear which of the two main programs (France 1 and 2, both originating in Paris) is more to blame. It has been advocated that deliberate experiments be done, say in Lille and Lyon, to investigate the matter. The local mayors might decide, by tossing coins each morning, which one of the two programs would be locally relayed during the day. Sufficient statistics would allow us to test hypotheses about the joint probability distribution for A conceptions in Lille and B in Lyon following exposure to programs $a(=1,2)$ and b, respectively:

$$\rho(A, B | a, b).$$

You might at first think it pointless to consider such a *joint* distribution, expecting it to separate trivially into independent factors:

$$\rho_1(A | a) \rho_2(B | b).$$

But a moment of reflection will convince that this will not be so. For example, the weather in the two towns is correlated, although imperfectly. On fine evenings people do not watch television. They walk in the parks, and are moved by the beauty of the trees, the monuments, and of one another. This is especially so on Sundays. Let λ denote, collectively, variables like temperature, humidity,..., day of the week, that might be relevant for Lille, and μ likewise for Lyon. Only when such relevant variables are held fixed can the distribution be expected to factorize:

$$\rho(A, B | a, b, \lambda, \mu) = \rho_1(A | a, \lambda) \rho_2(B | b, \mu). \tag{1}$$

Then

$$\rho(A, B | a, b) = \int\int d\lambda \, d\mu \, \sigma(\lambda, \mu) \rho_1(A | a, \lambda) \rho_2(B | b, \mu), \tag{2}$$

where σ is some probability distribution for temperatures, humidities,..., and days of the week.

Now surely it would be very remarkable if the choice of program in Lille proved to be a causal factor in Lyon, or if the choice of program in

Lyon proved to be a causal factor in Lille. It would be very remarkable, that is to say, if ρ_1 in 2 had to depend on b, or ρ_2 on a. But, according to quantum mechanics, situations presenting just such a dilemma can be contrived. Moreover the peculiar long-range influence in question seems to go faster than light.[1-3]

Avoiding internal details for the moment, consider just a long black box with three inputs and three outputs. The inputs are three on–off switches – a master switch in the middle and a switch at each end. The outputs are three corresponding printers. The one in the middle prints 'yes' or 'no' soon after the start of a run, and the others each print 'yes' or 'no' when it ends. While the switches are 'off' the box restores itself as far as possible to some given initial condition in preparation for a run. The master switch is then operated and left 'on' for a predetermined time T. At time $(T - \delta)$, each of the other switches may or may not – depending, for example, on random signals from independent radioactive sources external to the black box – be thrown to 'on' for a time δ. The length L of the box is such that

$$L/c \gg \delta,$$

where c is the velocity of light. So the operation of a switch at one end would not, according to Einstein, be relevant to the output at the other end.

We will consider only runs certified by a 'yes' from the middle printer, and not mention it any more. It just guarantees, as will be seen, that the internal process gets off to a good start. Let A (with values ± 1) denote the yes/no response of the left printer, and $B(\pm 1)$ likewise for the right printer. Let a ($= 1, 2$) denote whether or not the left switch is operated during the run, and b ($= 1, 2$) likewise for the right switch. With sufficient statistics we can test hypotheses about the joint probability of A and B given a and b:

$$\rho(A, B | a, b).$$

Consider, then, the hypothesis that A and B fluctuate independently when the relevant causal factors, at time $T - \delta - \varepsilon$ say, whatever they may be, are sufficiently well specified – as would be expected for conceptions in Lille and Lyon with specified weather, day of the week, television programs, and so on. That is, assume there are variables λ and some probability distribution σ such that (2) holds.

Now in fact this hypothesis is quite restrictive, for in the present case ($|A| = |B| = 1$) it implies (by some trivial manipulations) the Clauser–Holt–

Horne–Shimony inequality

$$|P(a,b) + P(a,b')| + |P(a',b) - P(a',b')| \leqslant 2, \tag{3}$$

where P is the mean value of the product AB:

$$P(a,b) = \sum_{\substack{A = \pm 1 \\ B = \pm 1}} AB\rho(A,B|a,b). \tag{4}$$

But, according to quantum mechanics, boxes can be constructed for which the left-hand side of (3) takes values up to $2\sqrt{2}$. The difficulty would not arise, for (3) would not follow, if ρ_1 in (2) were allowed to depend on b, or ρ_2 on a. Such a dependence would not only be of mysteriously long range, but also, for the case presented, would have to propagate faster than light. The correlations of quantum mechanics are not explicable in terms of local causes.

Going into the black box, we could find what is sketched (at the 'Gedanken' level) in Fig. 1. Only the centre and one end are drawn. The other end is the mirror image of the first. An oven provides a beam of suitable atoms in their $(j, P) = (0, +)$ ground states. A pulse of laser photons γ_{00} is activated (after a predetermined delay during which remote equipment is alerted) by the master switch. This excites some atoms to a certain $(1, -)$ level (Fig. 2). Most of these decay straight back to the ground state, but some cascade back with emission of photons $\gamma_0, \gamma_1, \gamma_2$. Some such cases are identified by a γ_0 counter C_0 with a suitable filter. And, for some of these, photons γ_1 and γ_2 go towards detecting equipment at the two ends of the box. Filters F_1 and F_2 pass only the correct photons

Fig. 1. Centre and left-hand-side of Gedanken set-up.

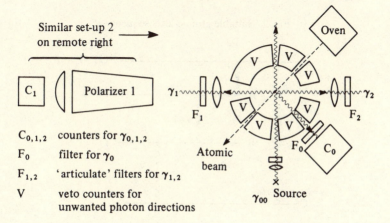

Similar set-up 2 on remote right →

$C_{0,1,2}$	counters for $\gamma_{0,1,2}$
F_0	filter for γ_0
$F_{1,2}$	'articulate' filters for $\gamma_{1,2}$
V	veto counters for unwanted photon directions

γ_1 and γ_2, and signal when they absorb wrong ones (i.e., they are a little more articulate than filters commercially available). Veto counters V identify events in which photons go off in other unwanted directions. Only the operation of counter C_0 and the nonoperation of the vetos V and $F_{1,2}$ authorize the middle printer to issue a 'yes' certificate for the event. Photons γ_1 and γ_2 then go towards distant detectors, C_1 and C_2, preceded by linear polarizers. These latter are set to pass polarizations at angles to the vertical controlled by the corresponding switches:

$$\phi_1 = (a-1)\pi/4, \quad \phi_2 = (b-3/2)\pi/4. \tag{5}$$

The firing or nonfiring of counters C_1 and C_2 authorizes the corresponding printers to print 'yes' or 'no'.

The heart of the matter is a strong correlation of polarization between photons γ_1 and γ_2, dictated by the spins and parities of levels A and C in Fig. 2. Because the atom has initially and finally no angular momentum, the photons can carry none away. For back-to-back photons this means a perfect circular polarization correlation – left-handed polarization for γ_1 implies left-handed γ_2, and right-handed γ_1 implies right-handed γ_2. Allowing also for parity conservation this translates into an equally strong linear polarization correlation: a given linear polarization on one side implies the same polarization on the other. In detail, in the ideal case of small opening angles and fully efficient counters, the probabilities of the various responses of C_1 and C_2 according to quantum mechanics are

$$\left.\begin{array}{l} \rho(\text{yes, yes}) = \rho(\text{no, no}) = \tfrac{1}{2}|\cos(\phi_1 - \phi_2)|^2 \\ \rho(\text{yes, no}) = \rho(\text{no, yes}) = \tfrac{1}{2}|\sin(\phi_1 - \phi_2)|^2, \end{array}\right\} \tag{6}$$

Fig. 2. Suitable atomic-level sequence.

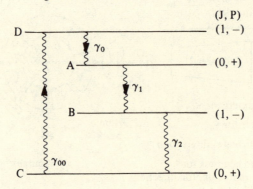

whence

$$P(a,b) = \cos 2(\phi_1 - \phi_2)$$

with (5), which gives $2\sqrt{2}$ for the left-hand side of (3), taking $a = b = 1$, $a' = b' = 2$.

Does nature really respect these remarkable predictions? A number of experiments have been done, on atomic cascades and other processes exhibiting similar correlations. The consensus is that quantum-mechanical predictions are well verified, and to very much better than a factor of $\sqrt{2}$.

The reservation must be made that all these experiments are very far in some respects – some more important than others – from the Gedanken ideal. For example, the photon counters are very inefficient. So 'no' is the normal and not very significant response at C_1 and C_2. Then $P = (1 - \Delta)$, where Δ is small and weakly dependent on a and b, so that the inequality in (3) is trivially satisfied. In addition, the real experiments have imperfect geometry. They do not have veto counters V, nor authorization counters C_0, nor 'articulate' filters F. And they are not done with one pair at a time, but look rather for (C_1, C_2) coincidences with a continuous source. What is verified in these experiments then is essentially that the coincidence rate for C_1 and C_2 – proportional to ρ (yes, yes) of (6) – is rather closely that predicted by quantum mechanics, when source strength, geometry and various inefficiencies are allowed for in conventional ways.

It is difficult for me to believe that quantum mechanics, working very well for currently practical set-ups, will nevertheless fail badly with improvements in counter efficiency and other factors just listed. However, there is at least one step towards the ideal which I am keen to see. So far, the polarizers have *not* been switched during the flight of the photons, but left in one setting or another for long periods. Such experiments can indicate an already remarkable influence of the polarizer setting on one side on the response of the counter on the other side. But plenty of time is left for this obscure influence to propagate across the equipment with subluminal velocity. For me it is important that Aspect[4] will effectively switch polarizer setting during the flight of the photons. It is difficult to rotate massive polarizers in nanoseconds. So he will have *two* polarizers on each side, preset at different angles, and rapidly reversible photon deflectors which can select one channel or another.

Let us anticipate that quantum mechanics works also for Aspect. How do we stand? I will list four of the attitudes that could be adopted.

(1) The inefficiencies of the counter, and so on, are essential. Quantum mechanics will fail in sufficiently critical experiments.

(2) There *are* influences going faster than light, even if we cannot control them for practical telegraphy. Einstein local causality fails, and we must live with this.

(3) The quantities *a* and *b* are not independently variable as we supposed. Whether apparently chosen by apparently independent radioactive devices, or by apparently separate Swiss National Lottery machines, or even by different apparently free-willed experimental physicists, they are in fact correlated with the same causal factors (λ, μ) as the *A* and *B*. Then Einstein local causality can survive. But apparently separate parts of the world become deeply entangled, and our apparent free will is entangled with them.

(4) The whole analysis can be ignored. The lesson of quantum mechanics is not to look behind the predictions of the formalism. As for the correlations, well, that's quantum mechanics. Just as the French legislators might shrug off a correlation between Lille and Lyon with, 'Well, that's people.'

Acknowledgements

This Comment is based on an invited talk at the Conference of the European Group for Atomic Spectroscopy, Orsay-Paris, 10–13 July 1979.

References

1 J. F. Clauser and A. Shimony, *Rep. Progr. Phys.* **41**, 1881 (1978) (a comprehensive review).

2 F. M. Pipkin, in *Advances in Atomic and Molecular Physics*, edited by D. R. Bates and B. Bederson. Academic Press, New York (1978), **14**, p. 281 (a comprehensive review).

3 B. d'Espagnat, *Scientific American* **241**, 158 (1979) (an extended introduction).

4 A. Aspect. *Phys. Rev.* **D14**, 1944 (1976).

14

de Broglie–Bohm, delayed-choice double-slit experiment, and density matrix

I will try to interest you in the de Broglie[1]–Bohm[2] version of non-relativistic quantum mechanics. It is, in my opinion, very instructive. It is experimentally equivalent to the usual version insofar as the latter is unambiguous. But it does not require, in its very formulation, a vague division of the world into 'system' and 'apparatus,' nor of history into 'measurement' and 'nonmeasurement.' So it applies to the world at large, and not just to idealized laboratory procedures. Indeed the de Broglie–Bohm theory is sharp where the usual one is fuzzy, and general where the usual one is special. This is not a systematic exposition[3], but only an illustration of the ideas with a particularly nice example, and then some remarks on the role of the density matrix – in tribute to the title of this conference.

No one more eloquently than John A. Wheeler[4] has presented the delayed-choice double slit experiment. A pulsed particle source S (see Fig. 1) is so feeble that not more than one particle is emitted per pulse. The associated wave pulse B falls on a screen with slits 1 and 2. The

Fig. 1. A de Broglie wave pulse B from a particle source S traverses a screen with slits 1 and 2. The waves B′ emerging from the slits are focussed by lenses on particle counters C_1 and C_2. A photographic plate P may or may not be pushed into the interference region.

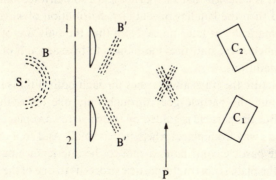

transmitted pulses B′ are focussed by off-centred lenses into intersecting plane wave trains which fall finally on particle counters C_1 and C_2 – unless a photographic plate P is pushed into the region where the two wave trains interpenetrate. The decision, to interpose the plate or not, is made only *after* the pulse has passed the slits. As a result of this choice the particle *either* falls on one of the two counters, indicating passage through one of the two slits, *or* contributes one of the spots on the photographic plate building an interference pattern after many repetitions. Sometimes the interference pattern is held to imply 'passage of the particle through both slits' – in some sense. Here it seems possible to *choose, later,* whether the particle, *earlier,* passed through one slit or two! Perhaps it is better not to think about it. 'No phenomenon is a phenomenon until it is an observed phenomenon.'

Consider now the de Broglie–Bohm version. To the question 'wave or particle?' they answer 'wave *and* particle.' The wave $\psi(t, \mathbf{r})$ is that of wave mechanics – but conceived, in the tradition of Maxwell and Einstein, as an objective field, and not just as some 'ghost wave' of information (of some presumably well-informed observer?). The particle rides along on the wave at some position $\mathbf{x}(t)$ with velocity.

$$\dot{\mathbf{x}}(t) = \frac{1}{m} \frac{\partial}{\partial \mathbf{r}} \operatorname{Im} \log \psi(t, \mathbf{r})|_{\mathbf{r}=\mathbf{x}} \tag{1}$$

This equation has the property that a probability distribution for $\mathbf{x}$ at time t

$$d^3\mathbf{x} |\psi(t, \mathbf{x})|^2$$

evolves into a distribution

$$d^3\mathbf{x} |\psi(t', \mathbf{x})|^2$$

at time t'. It is *assumed* that the particles are so delivered initially by the source, and then the familiar probability distribution of wave mechanics holds automatically at later times. Note that the only use of probability here is, as in classical statistical mechanics, to take account of uncertainty in initial conditions.

In this picture the wave always goes through both slits (as is the nature of waves) and the particle goes through only one (as is the nature of particles). But the particle is guided by the wave toward places where $|\psi|^2$ is large, and away from places where $|\psi|^2$ is small. And so if the plate is in position the particle contributes a spot to the interference pattern on the plate, or if the plate is absent the particle proceeds to one of the counters. In neither case is the earlier motion, of either particle or wave, affected by the later insertion or noninsertion of the plate. Clearly the particle pursues a bent path in the region where the wave trains interpenetrate[7]. It is vital here to put away the classical prejudice that a particle moves on a straight path in 'field-free' space – free, that is, from fields other than the de Broglie–Bohm! Indeed (in the absence of the plate) a particle passing through slit 1 falls finally not on counter C_1 but on C_2, and vice versa! It is clear just by symmetry that on the symmetry plane the perpendicular component of $\dot{x}$ vanishes. The particle does not cross this plane. The naive classical picture has the particle, arriving on a given counter, going through the wrong slit.

Suppose next that detectors are added to the setup just behind slits 1 and 2 to register the passage of the particles. If we wish to follow the story after these detectors have or have not registered we cannot pretend that they are passive external devices (as we did for screen and lenses). They have to be included in the system. Consider then an initial wavefunction

$$\Psi(0) = \psi(0,\mathbf{r})D_1^0(0,\mathbf{r}_1,\ldots)D_2^0(0,\mathbf{r}_2,\ldots)$$

where D_1^0 and D_2^0 are many-body wavefunctions for undischarged counters. Solution of the many-body Schrödinger equation yields a wavefunction

$$\Psi(t) = \Psi_1(t) + \Psi_2(t)$$
$$\left.\begin{aligned}\Psi_1(t) &= \psi_1(t,\mathbf{r})D_1^1(t,\mathbf{r}_1,\ldots)D_2^0(t,\mathbf{r}_2,\ldots)\\ \Psi_2(t) &= \psi_2(t,\mathbf{r})D_1^0(t,\mathbf{r}_1,\ldots)D_2^1(t,\mathbf{r}_2,\ldots)\end{aligned}\right\} \qquad (2)$$

where the ψs are the two plane wave trains and the D^1s are wavefunctions for discharged counters. Let us suppose that a discharged counter pops up a flag saying 'yes' just to emphasize that it is a macroscopically different thing from an undischarged counter, in a very different region of configuration space.

The many-particle generalization of (1) gives for the particle of particular interest

$$\dot{\mathbf{x}}(t) = \frac{1}{m}\frac{\partial}{\partial \mathbf{r}}\,\mathrm{Im}\log\Psi(t,\mathbf{r},\mathbf{r}_1,\mathbf{r}_2,\ldots)\left|\begin{smallmatrix}\mathbf{r}\;=\mathbf{x}(t)\\ \mathbf{r}_1=\mathbf{x}_1(t),\,\text{etc.}\\ \mathbf{r}_2=\mathbf{x}_2(t),\,\text{etc.}\end{smallmatrix}\right. \qquad (3)$$

Evaluation of this requires, in general, specification of not only $x(t)$ but also of the positions of all other particles. However, in the simple case of (2) the positions of other particles are sufficiently specified by 'detector 1 has

discharged' or 'detector 2 has discharged.' The configurations so described
are so different (grossly, macroscopically, so) that then only either Ψ_1 *or*
Ψ_2 is significantly different from zero. Moreover, since for either Ψ_1 or
Ψ_2 the variable $\mathbf{r}$ appears only in the factor ψ_1 or ψ_2, the complicated
formula (3) reduces to the simple formula (1):

$$\dot{\mathbf{x}} = \frac{1}{m}\frac{\partial}{\partial \mathbf{x}} \operatorname{Im} \log \psi_1(t,\mathbf{x}) = \mathbf{v}_1$$

or

$$\dot{\mathbf{x}} = \frac{1}{m}\frac{\partial}{\partial \mathbf{x}} \operatorname{Im} \log \psi_2(t,\mathbf{x}) = \mathbf{v}_2$$

where $\mathbf{v}_1$ and $\mathbf{v}_2$ are velocities associated with the two plane wave trains.

This reduction from Ψ_1 *and* Ψ_2 to Ψ_1 *or* Ψ_2, on partial (macroscopic)
specification of the configuration to be considered, illustrates the 'reduction
of the wavefunction' in the de Broglie–Bohm picture. It is a purely theoretical operation and one need not ask just when it happens and how long it
takes. The theorist does it when he finds it convenient.

The further reduction from Ψ_1 or Ψ_2 to ψ_1 or ψ_2 is a reduction from
many particles to a few (one in this case). It illustrates how with a partial
specification of the world at large it becomes possible in practice to deal
with a small quantum system – although in principle the correct
application of the theory is to the world as a whole. We made such a
reduction of the system tacitly in the beginning when we said that certain
screens and lenses, etc., were in position, but did not include them or the
world at large in the quantum system. Note that in the de Broglie–Bohm
scheme this singling out of a 'system' is a practical thing defined by
circumstances, and is not already in the fundamental formulation of the
theory.

Consider now the density matrix. When it is specified that counter 1
(say) has discharged, the conventional one-particle density matrix (with
disregard of trivial normalization factors) is

$$\rho(\mathbf{x},\mathbf{x}') = \psi_1(\mathbf{x})\psi_1^*(\mathbf{x}')$$

and the velocity $\dot{\mathbf{x}}_1 = \mathbf{v}_1$ is given equally by (1) or

$$\dot{\mathbf{x}}_1 = \frac{1}{m}\operatorname{Im}\left\{ [\rho(\mathbf{x},\mathbf{x}')]^{-1}\frac{\partial}{\partial \mathbf{x}}\rho(\mathbf{x},\mathbf{x}') \right\}_{\mathbf{x}=\mathbf{x}'} \tag{4}$$

But this is a rather trivial case. When it is not specified which counter has discharged the conventional density matrix is

$$\rho(\mathbf{x}, \mathbf{x}') = \int d^3\mathbf{x}_1 d^3\mathbf{x}_2 \dots \Psi(\mathbf{x}, \mathbf{x}_1, \mathbf{x}_2, \dots)\Psi^*(\mathbf{x}', \mathbf{x}_1, \mathbf{x}_2 \dots)$$

$$= \psi_1(\mathbf{x})\psi_1^*(\mathbf{x}') + \psi_2(\mathbf{x})\psi_2^*(\mathbf{x}')$$

I do not see how to recover from this the fact that we have (nearly always) velocity either $\mathbf{v}_1$ or $\mathbf{v}_2$. Naive application of (4) gives something else. So in the de Broglie–Bohm theory a fundamental significance is given to the wavefunction, and it cannot be transferred to the density matrix. This is here illustrated for the one-particle density matrix, but it equally so for the world density matrix if a probability distribution over world wavefunctions is considered. Of course the density matrix retains all its usual practical utility in connection with quantum statistics.

That the above treatment of the detectors was greatly oversimplified does not affect the main points made. Real detectors would respond in a variety of ways to particles traversing in a variety of ways. Not only would ψ_1 and ψ_2 become incoherent, but each would be replaced by many incoherent parts.

That the theory is supposed to apply fundamentally to the world as a whole requires ultimately that any 'observers' be included in the system. This raises no particular problem so long as they are conceived as not essentially different from computers, equipped perhaps with 'random' number generators. Then everything is in fact predetermined at the fundamental level – including the 'late' decision whether to insert the plate. To include creatures with genuine free will would require some development, and here the de Broglie–Bohm version might develop differently from the usual approach. To make the issue experimental would require identification of situations in which the differences betweeen computers and free agents were essential.

That the guiding wave, in the general case, propagates not in ordinary three-space but in a multidimensional-configuration space is the origin of the notorious 'nonlocality' of quantum mechanics[5]. It is a merit of the de Broglie–Bohm version to bring this out so explicitly that it cannot be ignored.[6]

References

1 L. de Broglie, *Tentative d'interpretation causale et non-linéaire de la méchanique ondulatoire*. Gauthier-Villars, Paris (1956).
2 D. Bohm, *Phys. Rev.* **85**, 166, 180 (1952).

3 J. S. Bell, CERN TH 1424 (1971).
4 J. A. Wheeler, in *Mathematical Foundation of Quantum Mechanics*, A. R. Marlow, Ed. Academic, New York (1978).
5 B. d'Espagnat, *Sci. Am.*, Nov. (1979).
6 J. S. Bell, *Rev. Mod. Phys.* **38**, 447 (1966).
7 C. Phillipidis, C. Dewdney, and B. J. Hiley, *Nuovo Cimento* **52B**, 15 (1979).

15

Quantum mechanics for cosmologists

1 Introduction

Cosmologists, even more than laboratory physicists, must find the usual interpretive rules of quantum mechanics a bit frustrating[1]:

> '...any result of a measurement of a real dynamical variable is one of its eigenvalues...'
> '...if the measurement of the observable... is made a large number of times the average of all the results obtained will be...'
> '...a measurement always causes the system to jump into an eigenstate of the dynamical variable that is being measured...'

It would seem that the theory is exclusively concerned with 'results of measurement' and has nothing to say about anything else. When the 'system' in question is the whole world where is the 'measurer' to be found? Inside, rather than outside, presumably. What exactly qualifies some subsystems to play this role? Was the world wave function waiting to jump for thousands of millions of years until a single-celled living creature appeared? Or did it have to wait a little longer for some more highly qualified measurer – with a Ph.D.? If the theory is to apply to anything but idealized laboratory operations, are we not obliged to admit that more or less 'measurement-like' processes are going on more or less all the time more or less everywhere? Is there ever then a moment when there is no jumping and the Schrödinger equation applies?

The concept of 'measurement' becomes so fuzzy on reflection that it is quite surprising to have it appearing in physical theory *at the most fundamental level.* Less surprising perhaps is that mathematicians, who need only simple axioms about otherwise undefined objects, have been able to write extensive works on quantum measurement theory – which experimental physicists do not find it necessary to read. Mathematics has been well called[2] 'the subject in which we never know what we are talking about'. Physicists, confronted with such questions, are soon making measurement a matter of degree, talking of 'good' measurements and 'bad' ones. But the postulates quoted above know nothing of 'good' and 'bad'. And does not any *analysis* of measurement require concepts more *fundamental* than measurement? And should not the fundamental theory be about these more fundamental concepts?

One line of development towards greater physical precision would be to have the 'jump' in the equations and not just the talk – so that it would come about as a dynamical process in dynamically defined conditions. The jump violates the linearity of the Schrödinger equation, so that the new equations (or equations) would be non-linear. It has been conjectured[3] that such non-linearity might be especially important precisely in connection with the functioning of conscious organisms – i.e., 'observers'.

It might also be that the non-linearity has nothing in particular to do with consciousness, but becomes important[4] for any large object, in such a way as to suppress superposition of macroscopically different states. This would be a mathematical realization of at least one version of the 'Copenhagen interpretation', in which large objects, and especially 'apparatus', must behave 'classically'. Cosmologists should note, by the way, that the suppression of such macroscopic superpositions is vital to Rosenfeld's notion[5] of an unquantized gravitational field – whose source (roughly speaking) would be the quantum expectation value of the energy density. If this were attempted with wave functions grossly ambiguous about, say, the relative positions of sun and planets, serious problems would quickly appear.

There have been several studies[6] of non-linear modifications of the Schrödinger equation. But none of these modifications (as far as I know) has the property required here, of having little impact for small systems but nevertheless suppressing macroscopic superpositions. It would be good to know how this could be done.

No more will be said in this paper about such hypothetical non-linearities. I will consider rather theories in which a linear Schrödinger equation is held to be exactly and universally correct. There is then no 'jumping', no 'reducing', no 'collapsing', of the wave function. Two such theories will be analyzed, one due to de Broglie[7] and Bohm[7] and the other to Everett[8]. It seems to me that the close relationship of the Everett theory to the de Broglie–Bohm theory has not been appreciated, and that as a result the really novel element in the Everett theory has not been identified. This really novel element, in my opinion, is a repudiation of the concept of the 'past', which could be considered in the same liberating tradition as Einstein's repudiation of absolute simultaneity.

It must be said that the version presented here might not be accepted by the authors cited. This is to be feared particularly in the case of Everett. His theory was for long completely obscure to me. The obscurity was lightened by the expositions of de Witt[8]. But I am not sure that my present understanding coincides with that of de Witt, or with that of Everett, or that a simultaneous coincidence with both would be possible[9].

The following starts with a review of some relevant aspects of conventional quantum mechanics, in terms of a simple particular application. The problems to which the unconventional versions are addressed are then stated in more detail, and finally the de Broglie–Bohm and Everett theories are formulated and compared.

2 Common ground

To illustrate some points which are not in question, before coming to some which are, let us look at a particular example of quantum mechanics in actual use. A nice example for our purpose is the theory of formation of an α particle track in a set of photographic plates. The essential ideas of the analysis have been around at least since 1929 when Mott[10] and Heisenberg[11] discussed the theory of Wilson cloud chamber tracks[12]. Yet somehow many students are left to rediscover for themselves ideas of this kind. When they do so it is often with a sense of revelation; this seems to be the origin of several published papers.

Let the α particle be incident normally on the stack of plates and excite various atoms or molecules in a way permitting development of blackened spots. In a first approach[11] to the problem only the α particle is considered as a quantum mechanical system, and the plates are thought of as external measuring equipment permitting a sequence of measurements of transverse position of the α particle. Associated with each such measurement there is a 'reduction of the wave packet' in which all of the incident de Broglie wave except that near the point of excitation is eliminated. If the 'position measurement' were of perfect precision the reduced wave would emerge in fact from a point source and, by ordinary diffraction theory, then spread over a large angle. However, the precision is presumably limited by something like the atomic diameter $a \approx 10^{-8}$ cm. Then the angular spread can be as little as

$$\triangle \theta \approx (ka)^{-1}$$

with an α particle of about one MeV, for example, $k \approx 10^{13}$ cm^{-1}, and with $a \approx 10^{-8}$ cm

$$\triangle \theta \approx 10^{-5} \text{ radians.}$$

In this way, one can understand that the sequence of excitations in the different plates approximate very well a straight line pointing to the source.

This first approach may seem very crude. Yet in an important sense it is an accurate model of all applications of quantum mechanics.

In a second approach we can regard the photographic plates also as part of the quantum mechanical system. As Heisenberg remarks 'this procedure is more complicated than the preceding method, but has the advantage that the discontinuous change in the probability function recedes one step and seems less in conflict with intuitive ideas'. To minimize the increased complication we will consider only highly simplified 'photographic plates'. They will be envisaged as zero temperature mono-atomic layers of atoms each with only one possible excited state, the latter supposed to be rather long-lived. Moreover, we will continue to neglect the possibility of scattering without excitation – i.e., elastic scattering, which is not very realistic.

Suppose that the α particle originates in a long-lived radioactive source at position $\mathbf{r}_0$ and can be represented initially by the steady state wave function

$$\psi(\mathbf{r}) = \frac{e^{ik_0|\mathbf{r}-\mathbf{r}_0|}}{|\mathbf{r}-\mathbf{r}_0|}$$

Let ϕ_0 denote the ground state of the stack of plates. Let $n(=1,2,3,\ldots)$ enumerate the atoms of the stack and let

$$\phi(n_1, n_2, n_3, \ldots)$$

denote a state of the stack in which atoms $n_1, n_2, n_3, \ldots$ are excited. In the absence of α particle stack interaction the combined state would be simply

$$\phi_0 \frac{e^{ik_0|\mathbf{r}-\mathbf{r}_0|}}{|\mathbf{r}-\mathbf{r}_0|}$$

To this must be added, because of the interaction, scattered waves determined by solution of the many-body Schrödinger equation. In a conventional multiple scattering approximation the scattered waves are

$$\sum_N \sum_{n_1,n_2,\ldots n_N} \phi(n_1, n_2, \ldots n_N) \frac{e^{ik_N|\mathbf{r}-\mathbf{r}_N|}}{|\mathbf{r}-\mathbf{r}_N|} f_N(\theta_N)$$

$$\times \frac{e^{ik_{N-1}|\mathbf{r}_N-\mathbf{r}_{N-1}|}}{|\mathbf{r}_N-\mathbf{r}_{N-1}|} f_{N-1}(\theta_{N-1}) \times \cdots \frac{e^{ik_0|\mathbf{r}_1-\mathbf{r}_0|}}{|\mathbf{r}_1-\mathbf{r}_0|} \quad (1)$$

The general term here is a sum over all possible sequences of N atoms, with r_1 denoting the position of atom n_1, r_2 of atom n_2, and so on; $k_n = (k_{n-1}^2 - \varepsilon)^{1/2}$ where ε is a measure of atomic excitation energy; θ_n is the angle between $r_n - r_{n-1}$, and $r_{n+1} - r_n$ (or $r - r_N$ for $n = N$). Finally $f_n(\theta)$ is the inelastic scattering amplitude for an α particle of momentum k_{n-1} incident on a single atom; in the Born approximation for example we could give an explicit formula for $f(\theta)$ in terms of atomic wave functions, and would indeed find for it an angular spread

$$\triangle \theta \approx (ka)^{-1}$$

The relative probabilities for observing that various sequences of atoms n_1, n_2,... have been excited are given by the squares of the moduli of the coefficients of

$$\phi(n_1, n_2, \ldots)$$

It is again clear that because of the forward peaking of $f(\theta)$ excited sequences will form essentially straight lines pointing towards the source.

We considered here only the location, and not the timing, of excitations. If timing also had been observed then in the first kind of treatment the reduced wave after each excitation would have been an appropriate solution of the time-dependent Schrödinger equation, limited in extent in time as well as space. In the second kind of treatment some physical device for registering and recording times would have been included in the system. We will not go further into this here. The comparison between the first and second kinds of treatment would still be essentially along the following lines. But before coming to this comparison it will be useful later to have pointed out two of the several general features of quantum mechanics which are illustrated in the example just discussed.

The first concerns the mutual consistency of different records of the same phenomenon. In the stack of plates of the above example we have a sequence of 'photographs' of the α particle, and because the particle is not *too* greatly disturbed by the photographing, the sequence of records is fairly continuous. In this way, there is no difficulty for quantum mechanics in the continuity between successive frames of a movie film nor in the consistency between two movie films of the same phenomenon. Moreover, if instead of recording such information on a film, it is fed into the memory of a computer (which can incidentally be thought of as a model for the brain) there is no difficulty for quantum mechanics in the internal coherence of such a record – e.g., in the 'memory' that the α particle (or instrument pointer, or whatever) has passed through a sequence of adjacent positions. These are all just 'classical' aspects of the world which emerge from quantum mechanics at the appropriate level. They are called to attention here because later on we come to a theory which is fundamentally precisely about the contents of 'memories'.

The second point is the following. When the whole stack of plates is treated as a single quantum mechanical system, each α particle track is a single experimental result. To test the quantum mechanical probabilities requires then many such tracks. At the same time a *single* track, if sufficiently long, can be regarded as a collection of many independent *single* scattering events, which can be used to test the quantum mechanics of the single scattering process. That this is so is seen to emerge from the more complete treatment whenever interactions between plates are negligible (and when the energy loss ε is negligible). Of course, there could be statistical freaks, tracks with all scatterings up, or all down, etc., but the *typical* track, if long enough, will serve to test predictions for $|f(\theta)|^2$. The relevance of this remark is that later we are concerned with theories of the universe as a whole. Then there is no opportunity to repeat the experiment; history is given to us once only. We are in the position of having a single track, and it is important that the theory has still something to say – provided that this single track is not a freak, but a typical member of the hypothetical ensemble of universes that would exhibit the complete quantum distribution of tracks[13].

We return now to the comparison of the two kinds of treatment. The second treatment is clearly more serious than the first. But it is by no means final. Just as at first we supposed without analysis that the photographic plates could effect position measurements on the α particle, so we have now supposed without analysis the existence of equipment allowing the observation of atomic excitation. We can therefore contemplate a third treatment, and a fourth, and so on. Any natural end to this sequence is excluded by the very language of contemporary quantum theory, which never speaks of events in the system but only of the outcome of observations upon the system, implying always the existence of external equipment adapted to the observable in question. Thus the logical situation does not change in going from the first treatment to the second. Nor would it change on going further, although many people have been intimidated simply by increasing complexity into imagining that this might be so. In spite of its manifest crudity, therefore, we have to take quite seriously the first treatment above, as a faithful model of what we have to do in the end anyway.

It is therefore important to consider to what extent the first treatment is actually consistent with the second, and not simply superseded by the latter.

The consistency is in fact quite high, especially if we incorporate into the rather vaguely 'reduced' wave function of the first treatment the correct angular factor $f(\theta)$ from the second. Then the first method will give exactly the same distribution of excitations, and the same correlations between those in different plates. However, it must be stressed that this perfect agreement is only a result of idealizations that we have made, for example, the neglect of interactions between atoms (especially in different plates). To take accurate account of these we are simply obliged to adopt the second procedure, of regarding α particle and stack together as a single quantum mechanical system. The first kind of treatment would be manifestly absurd if we were concerned with an α particle incident on two atoms forming a single molecule. It is perhaps not absurd, but it is not exact, when we have 10^{23} atoms with somewhat larger spaces between. Therefore, the placing of the inevitable split, between quantum system and observing world, is not a matter of indifference.

So we go on displacing this Heisenberg split to include more and more of the world in the quantum system. Eventually we come to a level where the required observations are simply of macroscopic aspects of macroscopic bodies. For example, we have to observe instrument readings, or a camera may do the observing, then we may observe the photographs of the instrument readings, and so on. At this stage, we know very well from everyday experience that it does not matter whether we think of the camera as being in the system or in the observer – the transformation between the two points of view being trivial, because the relevant aspects of the camera are 'classical' and its reaction on the relevant aspects of the instrument negligible. Then at this level it becomes of no *practical* importance just where we put the Heisenberg split – provided of course that these 'classical' features of the macroscopic world emerge also from the quantum mechanical treatment. There is no reason to doubt that this is the case.

This is already illustrated in the example that we analyzed above. Thus the α particle is already largely 'classical' in its behaviour – preserving its identity, in a sense, as it is seen to move along a practically continuous and smooth path. Moreover, the different parts of the complete wave function (1) associated with different tracks can be to a considerable extent regarded as incoherent, as indicated by the success of the first kind of treatment. These 'classical' features can be expected to be still more pronounced for macroscopic bodies. The possibilities of seeing quantum interference phenomena are reduced not only by the shortness of de Broglie wavelength, which would make any such pattern extremely fine grained, but also by the tendency of such bodies to record their passage in the environment. With macroscopic bodies it is not necessary to ionize atoms; we have the steady radiation of heat for example, which would leave a 'track' even in the vacuum, and we have the excitation of the close packed low lying collective levels of both the body in question and neighbouring ones[14].

So there is no reason to doubt that the quantum mechanics of macroscopic objects yields an image of the familiar everyday world. Then the following rule for placing the Heisenberg split, although ambiguous in principle, is sufficiently unambiguous for practical purposes:

> *put sufficiently much into the quantum system that the inclusion of more would not significantly alter practical predictions.*

To ask whether such a recipe, however adequate in practice, is also a satisfactory formulation of fundamental physical theory, is to leave the common ground.

3 The problem

The problem is this: quantum mechanics is fundamentally about 'observations'. It necessarily divides the world into two parts, a part which is observed and a part which does the observing. The results depend in detail on just how this division is made, but no definite prescription for it is given. All that we have is a recipe which, because of practical human limitations, is sufficiently unambiguous for practical purposes. So we may ask with Stapp[15]: 'How can a theory which is *fundamentally* a procedure by which gross macroscopic creatures, such as human beings, calculate predicted probabilities of what they will observe under macroscopically specified circumstances ever be claimed to be a complete description of physical reality?'. Rosenfeld[16] makes the point with equal eloquence: '... the human observer, whom we have been at pains to keep out of the picture, seems irresistibly to intrude into it, since after all the macroscopic character of the measuring apparatus is imposed by the macroscopic structure of the sense organs and the brain. It thus looks as if the mode of description of quantum theory would indeed fall short of ideal perfection to the extent that it is cut to the measure of man'.

Actually these authors feel that the situation is acceptable. As indicated by the quotations, they are among the more thoughtful of those who do so. Stapp finds reconciliation in the pragmatic philosophy of William James. On this view, the situation in quantum mechanics is not peculiar. But rather the concepts of 'real' or 'complete' truth are quite generally mirages. The only legitimate notion of truth is 'what works'. And quantum mechanics certainly 'works'. Rosenfeld seems to take much the same position, preferring however to keep academic philosophy out of it: 'we are not facing here any deep philosophical issue, but the plain common sense fact that it takes a complicated brain to do theoretical physics'. That is to say, that theoretical physics is quite necessarily cut to the measure of theoretical physicists.

In my opinion, these views are too complacent. The pragmatic approach which they examplify has undoubtedly played an indispensable role in the evolution of contemporary physical theory. However, the notion of the 'real' truth, as distinct from a truth that is presently good enough for us, has also played a positive role in the history of science. Thus Copernicus found a more intelligible pattern by placing the sun rather than the earth at the centre of the solar system. I can well imagine a future phase in which this happens again, in which the world becomes more intelligible to human beings, even to theoretical physicists, when they do not imagine themselves to be at the centre of it.

Less thoughtful physicists sometimes dismiss the problem by remarking that it was just the same in classical mechanics. Now if this were so it would diminish classical mechanics rather than justify quantum mechanics. But actually, it is not so. Of course, it is true that also in classical mechanics any isolation of a particular system from the world as a whole involves approximation. But at least one can *envisage* an accurate theory, of the universe, to which the restricted account *is* an approximation. This is not possible in quantum mechanics, which refers always to an outside observer, and for which therefore the universe as a whole is an embarrassing concept. It could also be said (by one unduly influenced by positivistic philosophy) that even in classical mechanics the human observer is implicit, for what is interesting if not experienced? But even a human observer is no trouble (in principle) in classical theory – he can be included in the system (in a schematic way) by postulating a 'psycho-physical parallelism' – i.e., supposing his experience to be correlated with some functions of the co-ordinates. This is not possible in quantum mechanics, where some kind of observer is not only essential, but essentially outside. In classical mechanics we have a model of a theory which is not *intrinsically* inexact, for it neither needs nor is embarrassed by an observer.

Classical mechanics does, however, have the grave defect, as applied on the atomic scale, of not accounting for the data. For this good reason it has been abandoned on that scale. However, classical concepts have not thereby been expelled from physics. On the contrary, they remain essential on the 'macroscopic' scale, for[17] '...it is decisive to recognize that, however far the phenomena transcend the scope of classical physical explanation, the account of all evidence must be expressed in classical terms'. Thus contemporary theory employs both quantum wave functions ψ and classical variables x, and a description of any sufficiently large part of the world involves both:

$$(\psi, x_1, x_2, \ldots)$$

In our discussion of the α particle track, for example, implicit classical variables specified the position of the various plates, and the degrees of excitation of the atoms were also considered as classical variables for which probability distributions could be extracted from the calculations. In a more thorough treatment the degrees of excitation of atoms would be replaced as classical variables by the degrees of blackening of the developed plates. And so on. It seems natural to speculate that such a description might survive in a hypothetical accurate theory to which the contemporary recipe would be a working approximation. The ψs and xs would then presumably interact according to some definite equations. These would replace the rather vague contemporary 'reduction of the wave packet' – intervening at some ill-defined point in time, or at some ill-defined point in the analysis, with a lack of precision which, as has been said, is tolerable only because of human grossness.

Before coming to examples of such theories I would like to suggest two general principles which should, it seems to me, be respected in their construction. The first is that it should be possible to formulate them for small systems. If the concepts have no clear meaning for small systems it is likely that 'laws of large numbers' are being invoked at a fundamental level, so that the theory is fundamentally approximate. The second, related, point is that the concepts of 'measurement', or 'observation', or 'experiment', should not appear at a fundamental level. The theory should of course allow for particular physical set-ups, not very well defined as a class, having a special relationship to certain not very well-defined subsystems – experimenters. But these concepts appear to me to be too vague to appear at the base of a potentially exact theory. Thus the xs then would not be 'macroscopic' 'observables' as in the traditional theory, but some more fundamental and less ambiguous quantities – 'beables'[18].

The classical variables x were written just now as a discrete set. In relativistic theory continuous fields are likely to be more appropriate, in particular perhaps an energy density $T_{00}(t, \mathbf{x})$. In the following we consider only the nonrelativistic theory, with the particulate approximation

$$T_{00}(t, \mathbf{x}) = \sum m_n c^2 \delta(\mathbf{x} - \mathbf{x}_n(t))$$

This is parametrized by the finite set of all particle coordinates x_n.

4 The pilot wave

The duality indicated by the symbol

$$(\psi, x)$$

is a generalization of the original wave-particle duality of wave mechanics. The mathematics had to be done with waves ψ extending in space, and then had to be interpreted in terms of probabilities for localized events. At an early stage de Broglie[7] proposed a scheme in which particle and wave aspects were more closely integrated. This was reinvented in 1952 by Bohm[7]. Despite some curious features it remains, in my opinion, well worth attention as a model of what might be the logical structure of a quantum mechanics which is not intrinsically inexact.

To avoid arbitrary division of the world into system and apparatus, we must work straight away with some model of the world as a whole. Let this 'world' be simply a large number N of particles, with Hamiltonian

$$H = \sum_n \frac{\mathbf{p}_n^2}{2M_n} + \sum_{m>n} V_{mn}(\mathbf{r}_m - \mathbf{r}_n) \tag{2}$$

The world wave function $\psi(r,t)$, where r stands for all the rs, evolves according to

$$\frac{\partial}{\partial t}\psi(r,t) = -iH\psi \tag{3}$$

We will need the purely mathematical consequence of this that

$$\frac{\partial}{\partial t}\rho(r,t) + \sum_n \frac{\partial}{\partial \mathbf{r}_n}\cdot\mathbf{j}_n(r,t) = 0 \tag{4}$$

where

$$\rho(r,t) = |\psi(r,t)|^2 \tag{5}$$

$$\mathbf{j}_n(r,t) = M_n^{-1}\,\mathrm{Im}\left\{\psi^*(r,t)\frac{\partial}{\partial \mathbf{r}_n}\psi(r,t)\right\} \tag{6}$$

We have to add classical variables. A democratic way to do this is to add variables $x_1, x_2, \ldots x_N$ in one-to-one correspondence with the rs. The xs are supposed to have definite values at any time and to change according to

$$\frac{d}{dt}x_n = \mathbf{j}_n(x,t)/\rho(x,t) = \frac{1}{M_n}\frac{\partial}{\partial x_n}\,\mathrm{Im}\log\psi(x,t) \tag{7}$$

We then have a deterministic system in which everything is fixed by the initial values of the wave ψ and the particle configuration x. Note that in this compound dynamical system the wave is supposed to be just as 'real' and 'objective' as say the fields of classical Maxwell theory – although its action on the particles, (7), is rather original. *No one can understand this theory until he is willing to think of ψ as a real objective field rather than just a 'probability amplitude'. Even though it propagates not in 3-space but in 3N-space.*

From the 'microscopic' variables x can be constructed 'macroscopic' variables X

$$X_n = F_n(\mathbf{x}_1, \dots \mathbf{x}_N) \tag{8}$$

– including in particular instrument readings, image density on photographic plates, ink density on computer output, and so on. Of course, there is some ambiguity in defining such quantities – e.g., over precisely what volume should the discrete particle density be averaged to define the smooth macroscopic density? However, it is the merit of the theory that the ambiguity is not in the foundation, but only at the level of identifying objects of particular interest to macroscopic observers, and the ambiguity arises simply from the grossness of these creatures.

It is thus from the xs, rather than from ψ, that in this theory we suppose 'observables' to be constructed. It is in terms of the xs that we would define a 'psycho-physical parallelism' – if we were pressed to go so far. Thus it would be appropriate to refer to the xs as 'exposed variables' and to ψ as a 'hidden variable'. It is ironic that the traditional terminology is the reverse of this.

It remains to compare the pilot-wave theory with orthodox quantum mechanics at a practical level, which is that of the xs. A convenient device for this purpose is to imagine, in the context of the orthodox approach, a sort of ultimate observer, outside the world and from time to time observing its macroscopic aspects. He will see in particular other, internal, observers at work, will see what their instruments read, what their computers print out, and so on. In so far as ordinary quantum mechanics yields at the appropriate level a classical world, in which the boundary between system and observer can be rather freely moved, it will be sufficient to account for what such an ultimate observer would see. If he were to observe at time t a whole ensemble of worlds corresponding to an initial state

$$\psi(\mathbf{r}_1, \dots \mathbf{r}_N, 0)$$

he would see, according to the usual theory, a distribution of Xs given closely by

$$\rho(X_1, X_2, \ldots) = \int \mathrm{d}\mathbf{r}_1 \mathrm{d}\mathbf{r}_2 \cdots \mathrm{d}\mathbf{r}_N$$

$$\delta(X_1 - F_1(r))\delta(X_2 - F_2(r)) \cdots |\psi(r, t)|^2 \qquad (9)$$

with $\psi(t)$ obtained by solving the world Schrödinger equation. It would not be exactly this, for his own activities cause wave-packet reduction and spoil the Schrödinger equation. But macroscopic observations is supposed to have not much effect on subsequent macroscopic statistics. Thus (4) is closely the distribution implied by the usual theory. Moreover, it is easy to construct in the pilot-wave theory an ensemble of worlds which gives the distribution (9) exactly. It is sufficient that the configuration x should be distributed according to

$$\rho(\mathbf{x}, t)\mathrm{d}\mathbf{x}_1 \mathrm{d}\mathbf{x}_2 \cdots \mathrm{d}\mathbf{x}_N \qquad (10)$$

It is a consequence of (4) and (7) that (10) will hold at all times if it holds at some initial time. Thus it suffices to specify in the pilot-wave theory that the initial configuration x is chosen at random from an ensemble of configurations in which the distribution is $\rho(x, 0)$. It is only at this point, in defining a comparison class of possible initial worlds, that anything like the orthodox probability interpretation is invoked.

Then for instantaneous macroscopic configurations the pilot-wave theory gives the same distribution as the orthodox theory, insofar as the latter is unambiguous. However, this question arises: what is the good of *either* theory, giving distributions over a hypothetical ensemble (of worlds!) when we have only one world. The answer has been anticipated in the introductory discussion of the α particle track. A long track is on the one hand a single event, but is at the same time an ensemble of single scatterings. In the same way a single configuration of the world will show statistical distributions over its different parts. Suppose, for example, this world contains an actual ensemble of similar experimental set-ups. In the same way as for the α particle track it follows from the theory that the 'typical' world will approximately realize quantum mechanical distributions over such approximately independent components[13]. The role of the hypothetical ensemble is precisely to permit definition of the word 'typical'.

So much for instantaneous configurations. Both theories give also trajectories, by which instantaneous configurations at different times are linked up. In the traditional theory these trajectories, like the configurations, emerge only at the macroscopic level, and are constructed by successive wave-packet reduction. In the pilot-wave theory macroscopic trajectories are a consequence of the microscopic trajectories determined by the guiding formula (7).

To exhibit some features of these trajectories, consider a standard example from quantum measurement theory – the measurement of a spin component of a spin $\frac{1}{2}$ particle. A highly simplified model for this can be based on the interaction.

$$H = g(t)\sigma \frac{1}{\mathrm{i}} \frac{\partial}{\partial r} \tag{11}$$

where σ is the Pauli matrix for the chosen component and r is the 'instrument reading' coordinate. For simplicity take the masses associated with both particle and instrument reading to be infinite. Then other terms in the Hamiltonian can be neglected, and the time-dependent coupling $g(t)$ can be supposed to arise from the passage of the particle along a definite classical orbit through the instrument. Let the initial state be

$$\psi_m(0) = \phi(r)a_m \tag{12}$$

where $\phi(r)$ is a narrow wave packet centred on $r = 0$ and $m (= 1, 2)$ is a spin index; we choose the representation in which σ is diagonal. The solution of the Schrödinger equation

$$\frac{\partial \psi}{\partial t} = -\mathrm{i}H\psi$$

is

$$\psi_m(t) = \phi(r - (-1)^m h)a_m \tag{13}$$

where

$$h(t) = \int_{-\infty}^{t} \mathrm{d}t' g(t') \tag{14}$$

After a short time the two components of (13) will separate in r space. Observation of the instrument reading will then, in the traditional view, yield the values $+h$ or $-h$ with relative probabilities $|a_1|^2$ and $|a_2|^2$, and with small uncertainties given by the width of the initial wave packet. Because of wave-packet reduction, subsequent observation will reveal the instrument continuing along whichever of the two trajectories, $\pm h(t)$, was in fact selected.

Consider now the pilot-wave version. Nothing new has to be said about the orbital motion of the particle, which was already taken to be classical and fixed. We do now have a classical variable x for the instrument reading.

We could consider introducing classical variables for the spin motion, but in the simplest version[20] this is not done; instead the spin indices of the wave function are just summed over in constructing densities and currents

$$\rho(r,t) = \psi^*(r,t)\psi(r,t) \tag{15}$$

$$j(r,t) = \psi^*(r,t)g\sigma\psi(r,t) \tag{16}$$

with the summation implied, the slightly surprising form of j following from the gradient form of the coupling (11), and from the absence of the normal term (6) in the case of infinite mass. The motion of x is then determined by

$$\frac{dx}{dt} = j(x,t)/\rho(x,t)$$

or explicitly

$$\frac{dx}{dt} = g\frac{\sum_m' |a_m|^2 |\phi(x-(-1)^m h)|^2 (-1)^m}{\sum_m |a_m|^2 |\phi(x-(-1)^m h)|^2} \tag{17}$$

As soon as the wave packets have separated $\dot{x} = \pm g$, according to $x \approx \pm h$. Thus we have essentially the same two trajectories as the wave-packet reduction picture, and they will be realized with the same relative probabilities if x is supposed to have an initial probability distribution $|\phi(x)|^2$ – this is the familiar general consequence, for instantaneous configurations, of the method of construction. In any individual case, which trajectory is selected is actually determined by the initial x value. But when that value is not known, when it is known only to lie in the initial wave-packet, whether the particle is deflected up or down is indeterminate for practical purposes.

Consider now a slightly more complicated example, in which measurements of the above kind are made simultaneously on two spin $\frac{1}{2}$ particles. Denote by r_1 and r_2 the coordinates of the two instruments. If the initial state is

$$\psi_{mn}(0) = \phi(r_1)\phi(r_2)a_{mn}$$

solution of the Schrödinger equation yields

$$\psi_{mn}(t) = \phi(r_1-(-1)^m h_1)\phi(r_2-(-1)^n h_2)a_{mn} \tag{18}$$

with

$$h_1(t) = \int_{-\infty}^t dt' g_1(t'), \quad h_2(t) = \int_{-\infty}^t dt' g_2(t')$$

In the wave-packet reduction picture one of four possible trajectories, $(\pm h_1, \pm h_2)$, will be realized, the relative probabilities being given by $|a_{mn}|^2$. The pilot-wave picture will give again an account identical for practical purposes, although the outcome is in principle determined by initial values of variables x_1 and x_2.

But when examined in detail the microscopic trajectories are quite peculiar during the brief initial period in which the different terms in (18) still overlap in (r_1, r_2) space. The detailed time development of the xs is given by

$$
\left.
\begin{aligned}
\dot{x}_1 &= g_1 \dfrac{\displaystyle\sum_{m,n} (-1)^m |a_{mn}|^2 |\phi(x_1 - (-1)^m h_1)|^2 |\phi(x_2 - (-1)^n h_2)|^2}{\displaystyle\sum_{m,n} |a_{mn}|^2 |\phi(x_1 - (-1)^m h_1)|^2 |\phi(x_2 - (-1)^n h_2)|^2} \\[4mm]
\dot{x}_2 &= g_2 \dfrac{\displaystyle\sum_{m,n} (-1)^n |a_{mn}|^2 |\phi(x_1 - (-1)^m h_1)|^2 |\phi(x_2 - (-1)^n h_2)|^2}{\displaystyle\sum_{m,n} |a_{mn}|^2 |\phi(x_1 - (-1)^m h_1)|^2 |\phi(x_2 - (-1)^n h_2)|^2}
\end{aligned}
\right\}
\quad (19)
$$

These expressions simplify greatly when the two spin states are uncorrelated, i.e., when a_{mn} factorizes

$$ a_{mn} = b_m c_n $$

The factors referring to the second particle then cancel out in the expression for $\dot{x}_1$, and those referring to the first particle cancel in the expression for $\dot{x}_2$, so that we have just two independent motions of the instrument pointers of the type already discussed. However, in general the spin state does not factorize. One can even envisage situations in which the two particles interact at short range and strong spin correlations are induced which persist when the particles subsequently move far apart. Then it follows from (19) that the detailed behaviour of x_1 and x_2 depends not only on the programmes h_1 and h_2 respectively of the local instruments, but also on those of the remote instruments h_2 and h_1. The detailed dynamics is quite non-local in character.

Could it be that this strange non-locality is a peculiarity of the very particular de Broglie–Bohm construction of the classical sector, and could be removed by a more clever construction? I think not. It now seems[21] that the non-locality is deeply rooted in quantum mechanics itself and will persist in any completion. Could it be that in the context of relativistic quantum theory c would be a limiting velocity and the strange long-range effects would propagate only subluminally? Not so. The aspects of quantum mechanics demanding non-locality remain in relativistic quantum mechanics. It may well be that a relativistic version of the theory, while Lorentz invariant and local at the observational level, may be necessarily non-local and with a preferred frame (or aether) at the fundamental level[22]. Could we not then just omit this fundamental level and restrict the classical variables to some 'observable' 'macroscopic' level? The problem then would be to do this with clean mathematics, and not just talk.

It can be maintained that the de Broglie–Bohm orbits, so troublesome in this matter of locality, are not an essential part of the theory. Indeed it can be maintained that there is no need whatever to link successive configurations of the world into a continuous trajectory. Keeping the instantaneous configurations, but discarding the trajectory, is the essential (in my opinion) of the theory of Everett.

5 Everett (?)

The Everett (?) theory of this section will simply be the pilot-wave theory without trajectories. Thus instantaneous classical configurations x are supposed to exist, and to be distributed in the comparison class of possible worlds with probability $|\psi|^2$. But no pairing of configurations at different times, as would be effected by the existence of trajectories, is supposed. And it is pointed out that no such continuity between present and past configurations is required by experience.

I would really prefer to leave the formulation at that, and proceed to elucidate the last sentence. But some additional remarks must be made for readers of Everett and De Witt, who may not immediately recognize the formulation just made.

(A) First there is the 'many-universe' concept given prominence by Everett and De Witt. In the usual theory it is supposed that only one of the possible results of a measurement is actually realized on a given occasion, and the wave function is 'reduced' accordingly. But Everett introduced the idea that *all* possible outcomes are realized every time, each in a different edition of the universe, which is therefore continually multiplying to accommodate all possible outcomes of every measurement. The psycho-physical parallelism is supposed such that our representatives in a given 'branch' universe are aware only of what is going on in that branch. Now it seems to me that this multiplication of universes is extravagant, and serves no real purpose in the theory, and can simply be dropped without repercussions. So I see no reason to insist on this particular difference between the

Everett theory and the pilot-wave theory – where, although the *wave* is never reduced, only *one* set of values of the variables *x* is realized at any instant. Except that the wave is in configuration space, rather than ordinary three-space, the situation is the same as in Maxwell–Lorentz electron theory[23]. Nobody ever felt any discomfort because the field was supposed to exist and propagate even at points where there was no particle. To have multiplied universes, to realize all possible configurations of particles, would have seemed grotesque.

(B) Then it could be said that the classical variables *x* do not appear in Everett and De Witt. However, it is taken for granted there that meaningful reference can be made to experiments having yielded one result rather than another. So instrument readings, or the numbers on computer output, and things like that, are the classical variables of the theory. We have argued already against the appearance of such vague quantities at a fundamental level. There is always some ambiguity about an instrument reading; the pointer has some thickness and is subject to Brownian motion. The ink can smudge in computer output, and it is a matter of practical human judgement that one figure has been printed rather than another. These distinctions are unimportant in practice, but surely the theory should be more precise. It was for that reason that the hypothesis was made of fundamental variables *x*, from which instrument readings and so on can be constructed, so that only at the stage of this construction, of identifying what is of direct interest to gross creatures, does an inevitable and unimportant vagueness intrude. I suspect that Everett and De Witt wrote as if instrument readings were fundamental only in order to be intelligible to specialists in quantum measurement theory.

(C) Then there is the surprising contention of Everett and De Witt that the theory 'yields its own interpretation'. The hard core of this seems to be the assertion that the probability interpretation emerges without being assumed. In so far as this is true it is true also in the pilot-wave theory. In that theory our unique world is supposed to evolve in deterministic fashion from some definite initial state. However, to identify which features are details critically dependent on the initial conditions (like whether the first scattering is up or down in an α particle track) and which features are more general (like the distribution of scattering angles over the track as a whole) it seems necessary to envisage a comparison class. This class we took to be a hypothetical ensemble of initial configurations with distribution $|\psi|^2$.

In the same way Everett has to attach weights to the different branches of his multiple universe, and in the same way does so in proportion to the norms of the relevant parts of the wave function. Everett and De Witt seem to regard this choice as inevitable. I am unable to see why, although of course it is a perfectly reasonable choice with several nice properties.

(D) Finally there is the question of trajectories, or of the association of a particular present with a particular past. Both Everett and De Witt do indeed refer to the structure of the wave function as a 'tree', and a given branch of a tree can be traced down in a unique way to the trunk. In such a picture the future of a given branch would be uncertain, or multiple, but the past would not. But, if I understand correctly, this tree-like structure is only meant to refer to a temporary and rough way of looking at things, during the period that the initially unfilled locations in a memory are progressively filled, labelling the different branches of the tree only by the macroscopic-type variables describing the contents of the locations. When a more fundamental description is adopted there is no reason to believe that the theory is more asymmetric in time than classical statistical mechanics. There also apparent irreversibility can arise (e.g., the increase of entropy) when coarse-grained variables are used. Moreover, De Witt says '...every quantum transition taking place on every star, in every galaxie, in every remote corner of the universe is splitting our local world in myriads of copies of itself'. Thus De Witt seems to share our idea that the fundamental concepts of the theory should be meaningful on a microscopic level, and not only on some ill-defined macroscopic level. But at the microscopic level there is no such asymmetry in time as would be indicated by the existence of branching and non-existence of debranching. Thus the structure of the wave function is not fundamentally tree-like. It does not associate a particular branch at the present time with any particular branch in the past any more than with any particular branch in the future. Moreover, it even seems reasonable to regard the coalescence of previously different branches, and the resulting interference phenomena, as *the* characteristic feature of quantum mechanics. In this respect an accurate picture, which does not have any tree-like character, is the 'sum over all possible paths' of Feynman.

Thus in our interpretation of the Everett theory there is no association of the particular present with any particular past. And the essential claim is that this does not matter at all. For we have no access to the past. We have only our 'memories' and 'records'. But these memories and records are in fact *present* phenomena. The instantaneous configuration of the xs can include clusters which are markings in notebooks, or in computer memories, or in human memories. These memories can be of the initial conditions in experiments, among other things, and of the results of those experiments. The theory should account for the present correlations between these present phenomena. And in this respect we have seen it to agree with ordinary quantum mechanics, in so far as the latter is unambiguous.

The question of making a Lorentz invariant theory on these lines raises intriguing questions. For reality has been identified only at a single time. This seems to be as much so in the many universe version, as in the one universe version. In a Lorentz invariant theory would there be different realities corresponding to different ways of defining the time direction in the four-dimensional space[24]? Or if these various realities are to be seen as different aspects of one, and therefore correlated somehow, is this not falling back towards the notion of trajectory?

Everett's replacement of the past by memories is a radical solipsism – extending to the temporal dimension the replacement of everything outside my head by my impressions, of ordinary solipsism or positivism. Solipsism cannot be refuted. But if such a theory were taken seriously it would hardly be possible to take anything else seriously. So much for the social implications[25]. It is always interesting to find that solipsists and positivists, when they have children, have life insurance.

In conclusion it is perhaps interesting to recall another occasion when the presumed accuracy of a theory required that the existence of present historical records should not be taken to imply that any past had indeed occurred. The theory was that of the creation of the world[26] in 4004 BC. During the 18th century growing knowledge of the structure of the earth seemed to indicate a more lengthy evolution. But it was pointed out that God in 4004 BC would quite naturally have created a going concern. The trees would be created with annular rings, although the corresponding number of years had not elapsed. Adam and Eve would be fully grown, with fully grown teeth and hair[27]. The rocks would be typical rocks, some occurring in strata and bearing fossils – of creatures that had never lived. Anything else would not have been reasonable:[28]

> Si le monde n'eut été à la fois jeune et vieux, le grand, le sérieux, le moral, disparaissaient de la nature, car ces sentiments tiennent par essence aux choses antiques.... L'homme-roi naquit lui-même à

trente années, afin de s'accorder par sa majesté avec les antiques grandeurs de son nouvel empire, de même que sa compagne compta sans doute seize printemps, qu'elle n'avait pourtant point vécu, pour être en harmonie avec les fleurs, les oiseaux, l'innocence, les amours, et toute la jeune partie de l'univers.

Notes and references

1 P. A. M. Dirac, *The Principles of Quantum Mechanics*, 3rd Edition. Oxford University Press (1930).
2 B. Russell, *Mysticism and Logic*, p. 75. Penguin, London (1953).
3 E. P. Wigner, in *The Scientist Speculates*, R. Good, Ed. Heinemann, London (1962).
   For a still more central role for the observer, see, C. M. Patton and J. A. Wheeler, in *Quantum Gravity*, Eds. C. Isham, R. Penrose and D. Sciama. Oxford (1975).
4 G. Ludwig, in *Werner Heisenberg und die Physik unserer Zeit*. Vierweg, Braunschweig (1961).
5 L. Rosenfeld, *Nuclear Phys.* **40**, 353 (1963).
6 L. de Broglie, *Nonlinear Wavemechanics*. Elsevier, Amsterdam (1960); B. Laurent and M. Roos, *Nuovo Cimento* **40**, 788 (1965); I. R. Shapiro, *Sov. J. Nucl. Phys.* **16**, 727 (1973); M. S. Marinov, *Sov. J. Nucl. Phys.* **19**, 173 (1974); M. Kupczynski, *Lett. Nuovo Cimento* **9**, ser. 2 no. 4, 134 (1974); B. Mielnik, *Comm. Math. Phys.* **37**, 221 (1974); P. Pearle, *Phys. Rev.* **D13**, 857 (1976); I. Bialnicki-Birula and J. Mycielski, *Ann. Phys.* **100**, 62 (1976); A. Shimony, *Phys. Rev.* **A20**, 394 (1979); T. W. B. Kibble, *Comm. Math. Phys.* **64**, 73 (1978); **65**, 189 (1979); T. W. B. Kibble and S. Randjbar-Daemi, *J. Phys.* **A13**, 141 (1980).
7 L. de Broglie, *Tentative d'Interpretation Causale et Non-linéaire de la Mécanique Ondulatoire*. Gauthier-Villars, Paris (1956); D. Bohm, *Phys. Rev.* **85**, 166, 180 (1952).
8 H. Everett, *Rev. Mod. Phys.* **29**, 454 (1957); J. A. Wheeler, *Rev. Mod. Phys.* **29**, 463 (1957); B. S. De Witt, *Physics Today* **23**, No. 9, p. 30 (1970); B. S. De Witt, in Proc. Int. School of Physics 'Enrico Fermi', Course IL: *Foundations of Quantum Mechanics*, B. d'Espagnat, Ed. Benjamin, New York (1971); L. N. Cooper and D. Van Vechten, *Am. J. Phys.* **37**, 1212 (1969). These five papers, a longer exposition by Everett, and a related paper by N. Graham [13], are collected in *The Many-Worlds Interpretation of Quantum Mechanics*, Ed., B. S. De Witt and N. Graham. Princeton, N.J., (1973).
   See also:
   B. S. De Witt and others, *Physics Today* **24**, No. 4, p. 36 (1971); J. S. Bell, in *Quantum Mechanics, Determinism, Causality and Particles*, Ed., M. Flato et al. Reidel, Dordrecht (1976).
9 In particular it is not clear to me that Everett and De Witt conceive in the same way the division of the wave function into 'branches'. For De Witt this division seems to be rather definite, involving a specific (although not very clearly specified) choice of variables (instrument readings) to have definite values in each branch. This choice is in no way dictated by the wave function itself (and it is only after it is made that the wave function becomes a complete description of De Witt's physical reality). Everett on the other hand (at least in some passages) seems to insist on the significance of assigning an arbitrarily chosen state to an arbitrarily chosen subsystem and evaluating the 'relative state' of the remainder. It is when arbitrary mathematical possibilities are given equal status in this way that

it becomes obscure to me that any physical interpretation has either emerged from, or been imposed on, the mathematics.

10 N. F. Mott, *Proc. Roy. Soc.* **A126**, 79 (1929).

11 W. Heisenberg, *Physical Principles of the Quantum Theory*. Chicago, (1930).

12 The particularly instructive nature of this example has been stressed by E. P. Wigner.

13 For elaboration of this point, see Everett, De Witt[8]; J. B. Hartle, *Am. J. Phys.* **36**, 704 (1968), and N. Graham[8].

14 The high probability of exciting collective levels is emphasized by H. D. Zeh, *Foundations of Physics* **1**, 69 (1970).

15 H. Stapp, UCRL-20294 (circa 1970). For later ideas of Stapp, see 21.

16 L. Rosenfeld, *Suppl. Prog. Theo. Phys.*, extra number 222 (1965).

17 N. Bohr, Discussion with Einstein, in *Albert Einstein*, Ed. P. A. Schilpp. Tudor, New York (1949).

18 J. S. Bell, in *The Physicists' Conception of Nature*, Ed. J. Mehra. Reidel, Dordrecht (1973).

19 There is a problem with (7) where ρ vanishes. A cheap way of avoiding it is to replace ρ and j in (4), (7), and (10), by $\bar{\rho}$ and $\bar{j}$, obtained from ρ and j by folding with a narrow Gaussian distribution in the $(r_1, r_2, \dots)$ space. Then $\bar{\rho}$ is always positive, while (4) remains valid. The deBB theory then gives $\bar{\rho}$ rather than ρ as probability distribution, but with sufficiently narrow Gaussian spread the difference is unimportant.

20 J. S. Bell, *Rev. Mod. Phys.* **38**, 447 (1966).

21 This question has been much discussed, and there has been an experimental programme to test the relevant aspects of quantum mechanics. Some papers, with many references, are: J. F. Clauser and A. Shimony, *Rep. Prog. Phys.* **41**, 1881 (1978); F. M. Pipkin, *Ann. Rev. Nuc. Sc.*, 1978; B. d'Espagnat, *Scientific American*, November 1978; H. Stapp, *Foundations of Physics*, **9**, 1–26 (1979); J. S. Bell, CERN, *Comments on Atomic and Molecular Physics*, **9**, 121–6 (1979).

22 P. H. Eberhard, *Nuovo Cimento* **46B**, 392 (1978).

23 But the following difference of detail is notable. In the Maxwell–Lorentz electron theory particles and field interacted in a reciprocal way. In the pilot-wave theory the wave influences the particles but is not influenced by them. Finding this peculiar, de Broglie,[7] always regarded the pilot-wave theory as just a stepping-stone on the way towards a more serious theory which would be in appropriate circumstances experimentally distinct from ordinary quantum mechanics.

24 Or would it be necessary to restrict memories to the here as well as the now? Point-sized reminiscers? See, H. D. Zeh, Foundations of Physics, **9**, 803–18 (1979).

25 The present work has much in common with an unpublished paper (CERN TH. 1424) presented at the International Colloquium on Issues in Contemporary Physics and Philosophy of Science, and their Relevance for our Society, Penn. State University, September 1971.

26 At 6 o'clock in the evening on October 22nd. J. Ussher, *Chronologia Sacra*. Oxford (1660).

27 They would have navels, although they had not been born. P. H. Gosse, *Omphalos* (1857).

28 F. de Chateaubriand, *Génie du Christianisme* (1802).

16

Bertlmann's socks and the nature of reality

Introduction

The philosopher in the street, who has not suffered a course in quantum mechanics, is quite unimpressed by Einstein–Podolsky–Rosen correlations[1]. He can point to many examples of similar correlations in everyday life. The case of Bertlmann's socks is often cited. Dr. Bertlmann likes to wear two socks of different colours. Which colour he will have on a given foot on a given day is quite unpredictable. But when you see (Fig. 1) that the first sock is pink you can be already sure that the second sock will not be pink. Observation of the first, and experience of Bertlmann, gives immediate information about the second. There is no accounting for tastes, but apart from that there is no mystery here. And is not the EPR business just the same?

Consider for example the particular EPR gedanken experiment of Bohm[2] (Fig. 2). Two suitable particles, suitably prepared (in the 'singlet spin state'), are directed from a common source towards two widely separated magnets followed by detecting screens. Each time the experiment is performed each of the two particles is deflected either up or down at

Fig. 1.

Les chaussettes
de M. Bertlmann
et la nature
de la réalité

Fondation Hugot
juin 17 1980

pink → → not pink

the corresponding magnet. Whether either particle separately goes up or down on a given occasion is quite unpredictable. But when one particle goes up the other always goes down and vice-versa. After a little experience it is enough to look at one side to know also about the other.

So what? Do we not simply infer that the particles have properties of some kind, detected somehow by the magnets, chosen à la Bertlmann by the source – differently for the two particles? Is it possible to see this simple business as obscure and mysterious? We must try.

To this end it is useful to know how physicists tend to think intuitively of particles with 'spin', for it is with such particles that we are concerned. In a crude classical picture it is envisaged that some internal motion gives the particle an angular momentum about some axis, and at the same time generates a magnetization along that axis. The particle is then like a little spinning magnet with north and south poles lying on the axis of rotation. When a magnetic field is applied to a magnet the north pole is pulled one way and the south pole is pulled the other way. If the field is uniform the net force on the magnet is zero. But in a non-uniform field one pole is pulled more than the other and the magnet as a whole is pulled in the corresponding direction. The experiment in question involves such non-

Fig. 2. Einstein–Podolsky–Rosen–Bohm gedanken experiment with two spin $\frac{1}{2}$ particles and two Stern–Gerlach magnets.

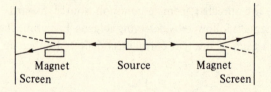

Magnet   Source   Magnet
Screen   Screen

Fig. 3. Forces on magnet in non-uniform magnetic field. The field points towards the top of the page and increases in strength in that direction.

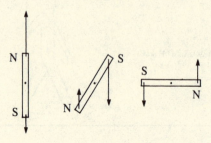

uniform fields – set up by so-called 'Stern–Gerlach' magnets. Suppose that the magnetic field points up, and that the strength of the field increases in the upward direction. Then a particle with south–north axis pointing up would be pulled up (Fig. 3). One with axis pointing down would be pulled down. One with axis perpendicular to the field would pass through the field without deflection. And one oriented at an intermediate angle would be deflected to an intermediate degree. (All this is for a particle of zero electric charge; when a charged particle moves in a magnetic field there is an additional force which complicates the situation.)

A particle of given species is supposed to have a given magnetization. But because of the variable angle between particle axis and field there would still be a range of deflections possible in a given Stern–Gerlach magnet. It could be expected then that a succession of particles would make a pattern something like Fig. 4 on a detecting screen. But what is observed in the simplest case is more like Fig. 5, with two distinct groups of deflections (i.e., up or down) rather than a more or less continuous band. (This simplest case, with just two groups of deflections, is that of so-called 'spin-$\frac{1}{2}$' particles; for 'spin-j' particles there are $(2j + 1)$ groups).

The pattern of Fig. 5 is very hard to understand in naïve classical terms. It might be supposed for example that the magnetic field first pulls the little magnets into alignment with itself, like compass needles. But even if this were dynamically sound it would account for only one group of deflections. To account for the second group would require 'compass-needles' pointing in the wrong direction. And anyway it is not dynamically sound. The internal angular momentum, by gyroscopic action, should stabilize the angle between particle axis and magnetic field. Well then, could it not be that the source for some reason delivers particles with axes pointing just one way or the other and not in between? But this is

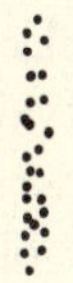

Fig. 4. Naive classical expectation for pattern on detecting screen behind Stern–Gerlach magnet.

Fig. 5. Quantum mechanical pattern on screen, with vertical Stern–Gerlach magnet.

easily tested by turning the Stern–Gerlach magnet. What we get (Fig. 6) is just the same split pattern as before, but turned around with the Stern–Gerlach magnet. To blame the absence of intermediate deflections on the source we would have to imagine that it anticipated somehow the orientation of the Stern–Gerlach magnet.

Phenomena of this kind[3] made physicists despair of finding any consistent space-time picture of what goes on on the atomic and subatomic scale. Making a virtue of necessity, and influenced by positivistic and instrumentalist philosophies[4], many came to hold not only that it is difficult to find a coherent picture but that it is wrong to look for one – if not actually immoral then certainly unprofessional. Going further still, some asserted that atomic and subatomic particles do not *have* any definite properties in advance of observation. There is nothing, that is to say, in the particles approaching the magnet, to distinguish those subsequently deflected up from those subsequently deflected down. Indeed even the particles are not really there.

For example[5], 'Bohr once declared when asked whether the quantum mechanical algorithm could be considered as somehow mirroring an underlying quantum reality: "There is no quantum world. There is only an abstract quantum mechanical description. It is wrong to think that the task of physics is to find out how Nature *is*. Physics concerns what we can say about Nature"'.

And for Heisenberg[6] '... in the experiments about atomic events we have to do with things and facts, with phenomena that are just as real as any phenomena in daily life. But the atoms or the elementary particles are not as real; they form a world of potentialities or possibilities rather than one of things or facts'.

And[7] 'Jordan declared, with emphasis, that observations not only *disturb* what has to be measured, they *produce* it. In a measurement of position, for example, as performed with the gamma ray microscope, "the electron is forced to a decision. We compel it *to assume a definite position*; previously it was, in general, neither here nor there; it had not yet made its decision for a definite position... If by another experiment the *velocity* of the electron is being measured, this means: the electron is compelled to decide itself for some exactly defined value of the velocity... we ourselves produce the results of measurement"'.

Fig. 6. Quantum mechanical pattern with rotated Stern–Gerlach magnet.

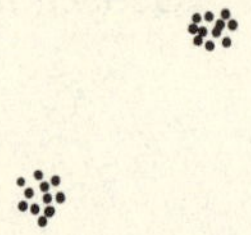

It is in the context of ideas like these that one must envisage the discussion of the Einstein–Podolsky–Rosen correlations. Then it is a little less unintelligible that the EPR paper caused such a fuss, and that the dust has not settled even now. It is as if we had come to deny the reality of Bertlmann's socks, or at least of their colours, when not looked at. And as if a child has asked: How come they always choose different colours when they *are* looked at? How does the second sock know what the first has done?

Paradox indeed! But for the others, not for EPR. EPR did not use the word 'paradox'. They were with the man in the street in this business. For them these correlations simply showed that the quantum theorists had been hasty in dismissing the reality of the microscopic world. In particular Jordan had been wrong in supposing that nothing was real or fixed in that world before observation. For after observing only one particle the result of subsequently observing the other (possibly at a very remote place) is immediately predictable. Could it be that the first observation somehow fixes what was unfixed, or makes real what was unreal, not only for the near particle but also for the remote one? For EPR that would be an unthinkable 'spooky action at a distance'[8]. To avoid such action at a distance they have to attribute, to the space-time regions in question, *real* properties in advance of observation, correlated properties, which *predetermine* the outcomes of these particular observations. Since these real properties, fixed in advance of observation, are not contained in quantum formalism[9], that formalism for EPR is *incomplete*. It may be correct, as far as it goes, but the usual quantum formalism cannot be the whole story.

It is important to note that to the limited degree to which *determinism* plays a role in the EPR argument, it is not assumed but *inferred*. What is held sacred is the principle of 'local causality' – or 'no action at a distance'. Of course, mere *correlation* between distant events does not by itself imply action at a distance, but only correlation between the signals reaching the two places. These signals, in the idealized example of Bohm, must be sufficient to *determine* whether the particles go up or down. For any residual undeterminism could only spoil the perfect correlation.

It is remarkably difficult to get this point across, that determinism is not a *presupposition* of the analysis. There is a widespread and erroneous conviction that for Einstein[10] determinism was always *the* sacred principle. The quotability of his famous 'God does not play dice' has not helped in this respect. Among those who had great difficulty in seeing Einstein's position was Born. Pauli tried to help him[11] in a letter of 1954:

...I was unable to recognize Einstein whenever you talked about him in either your letter or your manuscript. It seemed to me as if you had erected some dummy Einstein for yourself, which you then knocked down with great pomp. In particular Einstein does not consider the concept of 'determinism' to be as fundamental as it is frequently held to be (as he told me emphatically many times)...he *disputes* that he uses as a criterion for the admissibility of a theory the question: 'Is it rigorously deterministic?'...he was not at all annoyed with you, but only said you were a person who will not listen.

Born had particular difficulty with the Einstein–Podolsky–Rosen argument. Here is his summing up, long afterwards, when he edited the Born–Einstein correspondence[12].

The root of the difference between Einstein and me was the axiom that events which happen in different places A and B are independent of one another, in the sense that an observation on the state of affairs at B cannot teach us anything about the state of affairs at A.

Misunderstanding could hardly be more complete. Einstein had no difficulty accepting that affairs in different places could be correlated. What he could not accept was that an intervention at one place could *influence*, immediately, affairs at the other.

These references to Born are not meant to diminish one of the towering figures of modern physics. They are meant to illustrate the difficulty of putting aside preconceptions and listening to what is actually being said. They are meant to encourage *you*, dear listener, to listen a little harder.

Here, finally, is a summing-up by Einstein himself[13]:

If one asks what, irrespective of quantum mechanics, is characteristic of the world of ideas of physics, one is first of all struck by the following: the concepts of physics relate to a real outside world.... It is further characteristic of these physical objects that they are thought of as arranged in a space-time continuum. An essential aspect of this arrangement of things in physics is that they lay claim, at a certain time, to an existence independent of one another, provided these objects 'are situated in different parts of space'.

The following idea characterizes the relative independence of objects far apart in space (A and B): external influence on A has no direct influence on B...

There seems to me no doubt that those physicists who regard the descriptive methods of quantum mechanics as definitive in principle would react to this line of thought in the following way: they would drop the requirement... for the independent existence of the physical reality present in different parts of space; they would be justified in pointing out that the quantum theory nowhere makes explicit use of this requirement.

I admit this, but would point out: when I consider the physical phenomena known to me, and especially those which are being so successfully encompassed by quantum mechanics, I still cannot find any fact anywhere which would make it appear likely that (that) requirement will have to be abandoned.

I am therefore inclined to believe that the description of quantum mechanics... has to be regarded as an incomplete and indirect description of reality, to be replaced at some later date by a more complete and direct one.

2 Illustration

Let us illustrate the *possibility* of what Einstein had in mind in the context of the particular quantum mechanical predictions already cited for the EPRB gedanken experiment. These predictions make it hard to believe in the completeness of quantum formalism. But of course outside that formalism they make no difficulty whatever for the notion of local causality. To show this explicitly we exhibit a trivial *ad hoc* space-time picture of what might go on. It is a modification of the naive classical picture already described. Certainly something must be modified in that, to reproduce the quantum phenomena. Previously, we implicitly assumed for the net force in the direction of the field gradient (which we always take to be in the same direction as the field) a form

$$F \cos \theta \tag{1}$$

where θ is the angle between magnetic field (and field gradient) and particle axis. We change this to

$$F \cos \theta / |\cos \theta|. \tag{2}$$

Whereas previously the force varied over a continuous range with θ, it takes now just two values, $\pm F$, the sign being determined by whether the magnetic axis of the particle points more nearly in the direction of the field or in the opposite direction. No attempt is made to explain this change in the force law. It is just an *ad hoc* attempt to account for the observations. And of course it accounts immediately for the appearance of just two groups of particles, deflected either in the direction of the magnetic field or in the opposite direction. To account then for the

Einstein–Podolsky–Rosen–Bohm correlations we have only to assume that the two particles emitted by the source have oppositely directed magnetic axes. Then if the magnetic axis of one particle is more nearly along (than against) one Stern–Gerlach field, the magnetic axes of the other particle will be more nearly against (than along) a parallel Stern–Gerlach field. So when one particle is deflected up, the other is deflected down, and vice versa. There is nothing whatever problematic or mind-boggling about these correlations, with parallel Stern–Gerlach analyzers, from the Einsteinian point of view.

So far so good. But now go a little further than before, and consider *non*-parallel Stern–Gerlach magnets. Let the first be rotated away from some standard position, about the particle line of flight, by an angle a. Let the second be rotated likewise by an angle b. Then if the magnetic axis of either particle separately is randomly oriented, but if the axes of the particles of a given pair are always oppositely oriented, a short calculation gives for the probabilities of the various possible results, in the *ad hoc* model,

$$\left. \begin{array}{c} P(\text{up, up}) = P(\text{down, down}) = \dfrac{|a-b|}{2\pi} \\[2mm] P(\text{up, down}) = P(\text{down, up}) = \dfrac{1}{2} - \dfrac{|a-b|}{2\pi} \end{array} \right\} \quad (3)$$

where 'up' and 'down' are defined with respect to the magnetic fields of the two magnets. However, a quantum mechanical calculation gives

$$\left. \begin{array}{c} P(\text{up, up}) = P(\text{down, down}) = \dfrac{1}{2}\left(\sin\dfrac{a-b}{2}\right)^2 \\[2mm] P(\text{up, down}) = P(\text{down, up}) = \dfrac{1}{2} - \dfrac{1}{2}\left(\sin\dfrac{a-b}{2}\right)^2 \end{array} \right\} \quad (4)$$

Thus the *ad hoc* model does what is required of it (i.e., reproduces quantum mechanical results) only at $(a-b) = 0$, $(a-b) = \pi/2$ and $(a-b) = \pi$, but not at intermediate angles.

Of course this trivial model was just the first one we thought of, and it worked up to a point. Could we not be a little more clever, and devise a model which reproduces the quantum formulae completely? No. It cannot be done, so long as action at a distance is excluded. This point was realized only subsequently. Neither EPR nor their contemporary opponents were aware of it. Indeed the discussion was for long entirely concentrated on the points $|a-b| = 0$, $\pi/2$, and π.

3 Difficulty with locality

To explain this dénouement without mathematics I cannot do better than follow d'Espagnat[14,15]. Let us return to socks for a moment. One of the most important questions about a sock is 'will it wash'? A consumer research organization might make the question more precise: could the sock survive one thousand washing cycles at 45 °C? Or at 90 °C? Or at 0 °C? Then an adaptation of the Wigner–d'Espagnat inequality[16] applies. For any collection of new socks:

$$
\left.
\begin{array}{c}
\text{(the number that could pass at } 0° \text{ and not at } 45°) \\
\text{plus} \\
\text{(the number that could pass at } 45° \text{ and not at } 90°) \\
\text{is not less than} \\
\text{(the number that could pass at } 0° \text{ and not at } 90°)
\end{array}
\right\} \quad (5)
$$

This is trivial, for each member of the third group either could survive at 45°, and so is also in the second group, or could not survive at 45°, and so is also in the first group.

But trivialities like this, you will exclaim, are of no interest in consumer research! You are right; we are straining here a little the analogy between consumer research and quantum philosophy. Moreover, you will insist, the statement has no empirical content. There is no way of deciding that a given sock could survive at one temperature and not at another. If it did not survive the first test it would not be available for the second, and even if it did survive the first test it would no longer be new, and subsequent tests would not have the original significance.

Suppose, however, that the socks come in pairs. And suppose that we know by experience that there is little variation between the members of a pair, in that if one member passes a given test then the other also passes that same test *if* it is performed. Then from d' Espagnat's inequality we can infer the following:

$$
\left.
\begin{array}{c}
\text{(the number of pairs in which one could pass at } 0° \text{ and} \\
\text{the other not at } 45°) \\
\text{plus} \\
\text{(the number of pairs in which one could pass at } 45° \text{ and} \\
\text{the other not at } 90°) \\
\text{is not less than} \\
\text{(the number of pairs in which one could pass at } 0° \text{ and} \\
\text{the other not at } 90°)
\end{array}
\right\} \quad (6)
$$

This is not yet empirically testable, for although the two tests in each bracket are now on different socks, the different brackets involve different tests on the same sock. But we now add the random sampling hypothesis: if the sample of pairs is sufficiently large and if we choose at random a big enough subsample to suffer a given pair of tests, then the pass/fail fractions of the subsample can be extended to the whole sample with high probability. Identifying such fractions with *probabilities* in a thoroughly conventional way, we now have

$$
\left.
\begin{array}{c}
\text{(the probability of one sock passing at } 0° \text{ and} \\
\text{the other not at } 45°) \\
\text{plus} \\
\text{(the probability of one sock passing at } 45° \text{ and} \\
\text{the other not at } 90°) \\
\text{is not less than} \\
\text{(the probability of one sock passing at } 0° \text{ and} \\
\text{the other not at } 90°)
\end{array}
\right\} \quad (7)
$$

Moreover this is empirically meaningful is so far as probabilities can be determined by random sampling.

We formulated these considerations first for pairs of socks, moving with considerable confidence in those familiar objects. But why not reason similarly for the pairs of particles of the EPRB experiment? By blocking off the 'down' channels in the Stern–Gerlach magnets, allowing only particles deflected 'up' to pass, we effectively subject the particles to tests which they either pass or do not. Instead of temperatures we now have angles *a* and *b* at which the Stern–Gerlach magnets are set. The essential difference, a trivial one, is that the particles are paired à la Bertlmann – if one were to pass a given test the other would be sure to fail it. To allow for this we simply take the converse of the second term in each bracket:

$$
\left.
\begin{array}{c}
\text{(the probability of one particle passing at } 0° \text{ and} \\
\text{the other at } 45°) \\
\text{plus} \\
\text{(the probability of one particle passing at } 45° \text{ and} \\
\text{the other at } 90°) \\
\text{is not less than} \\
\text{(the probability of one particle passing at } 0° \text{ and} \\
\text{the other at } 90°)
\end{array}
\right\} \quad (8)
$$

In case any one finds the detour by socks a little long, let us look directly at this final result and see how trivial it is. We are assuming that particles have properties which dictate their ability to pass certain tests – whether or not these tests are in fact made. To account for the perfect anticorrelation when identical tests (parallel Stern–Gerlach magnets) are applied to the two members of a pair, we have to admit that the pairing is generalized à la Bertlmann – when one has the ability to pass a certain test, the other has not. Then the above assertion about pairs is equivalent to the following assertion about either member:

(the probability of being able to pass at 0° and
not able at 45°)

plus

(the probability of being able to pass at 45° and
not able at 90°)   (9)

is not less than

(the probability of being able to pass at 0° and
not able at 90°)

And this is indeed trivial. For a particle able to pass at 0° and not at 90° (and so contributing to the third probability in (9)) is either able to pass at 45° (and so contributes to the second probability) or not able to pass at 45° (and so contributes to the first probability).

However, trivial as it is, the inequality is not respected by quantum mechanical probabilities. From (4) the quantum mechanical probability for one particle to pass a magnet with orientation a and the other to pass a magnet with orientation b (called P (up, up)) in (4) is

$$\frac{1}{2}\left(\sin\frac{a-b}{2}\right)^2$$

Inequality (9) would then require

$$\tfrac{1}{2}(\sin 22.5°)^2 + \tfrac{1}{2}(\sin 22.5°)^2 \geqslant \tfrac{1}{2}(\sin 45°)^2$$

or

$$0.1464 \geqslant 0.2500$$

which is not true.

Let us summarize once again the logic that leads to the impasse. The EPRB correlations are such that the result of the experiment on one side immediately foretells that on the other, whenever the analyzers happen to be parallel. If we do not accept the intervention on one side as a causal influence on the other, we seem obliged to admit that the results on both sides are determined in advance anyway, independently of the intervention on the other side, by signals from the source and by the local magnet setting. But this has implications for non-parallel settings which conflict with those of quantum mechanics. So we *cannot* dismiss intervention on one side as a causal influence on the other.

It would be wrong to say 'Bohr wins again' (Appendix 1); the argument was not known to the opponents of Einstein, Podolsky and Rosen. But certainly Einstein could no longer write so easily, speaking of local causality '...I still cannot find any fact anywhere which would make it appear likely that that requirement will have to be abandoned'.

4 General argument

So far the presentation aimed at simplicity. Now the aim will be generality[17]. Let us first list some aspects of the simple presentation which are not essential and will be avoided.

The above argument relies very much on the perfection of the correlation (or rather anticorrelation) when the two magnets are aligned ($a = b$) and other conditions also are ideal. Although one could hope to approach this situation closely in practice, one could not hope to realize it completely. Some residual imperfection of the set-up would spoil the perfect anticorrelation, so that occasionally both particles would be deflected down, or both up. So in the more sophisticated argument we will avoid any hypothesis of perfection.

It was only in the context of perfect correlation (or anticorrelation) that *determinism* could be inferred for the relation of observation results to preexisting particle properties (for any indeterminism would have spoiled the correlation). Despite my insistence that the determinism was inferred rather than assumed, you might still suspect somehow that it is a preoccupation with determinism that creates the problem. Note well then that the following argument makes no mention whatever of determinism.

You might suspect that there is something specially peculiar about spin-$\frac{1}{2}$ particles. In fact there are many other ways of creating the troublesome correlations. So the following argument makes no reference to spin-$\frac{1}{2}$ particles, or any other particular particles.

Finally you might suspect that the very notion of particle, and particle orbit, freely used above in introducing the problem, has somehow led us astray. Indeed did not Einstein think that fields rather than particles are at the bottom of everything? So the following argument will not mention particles, nor indeed fields, nor any other particular picture of what goes on at the microscopic level. Nor will it involve any use of the words 'quantum mechanical system', which can have an unfortunate effect on the discussion. The difficulty is not created by any such picture or any such terminology. It is created by the predictions about the correlations in the visible outputs of certain conceivable experimental set-ups.

Consider the general experimental set-up of Fig. 7. To avoid inessential details it is represented just as a long box of unspecified equipment, with three inputs and three outputs. The outputs, above in the figure, can be three pieces of paper, each with either 'yes' or 'no' printed on it. The central input is just a 'go' signal which sets the experiment off at time t_1. Shortly after that the central output says 'yes' or 'no'. We are only interested in the 'yes's, which confirm that everything has got off to a good start (e.g., there are no 'particles' going in the wrong directions, and so on). At time $t_1 + T$ the other outputs appear, each with 'yes' or 'no' (depending for example on whether or not a signal has appeared on the 'up' side of a detecting screen behind a local Stern–Gerlach magnet). The apparatus then rests and recovers internally in preparation for a subsequent repetition of the experiment. But just before time $t_1 + T$, say at time $t_1 + T - \delta$, signals a and b are injected at the two ends. (They might for example dictate that Stern–Gerlach magnets be rotated by angles a and b away from some standard position). We can arrange that $c\delta \ll L$, where c is the velocity of light and L the length of the box; we would not then expect the signal at one end to have any influence on the output at the other, for lack of time, whatever hidden connections there might be between the two ends.

Sufficiently many repetitions of the experiment will allow tests of hypotheses about the joint conditional probability distribution

$$P(A, B | a, b)$$

for results A and B at the two ends for given signals a and b.

Now of course it would be no surprise to find that the two results A and B are correlated, i.e., that P does not split into a product of independent factors:

$$P(A, B | a, b) \neq P_1(A | a) P_2(B | b)$$

But we will argue that certain particular correlations, realizable according

Fig. 7. General EPR set-up, with three inputs below and three outputs above.

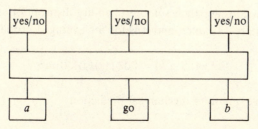

to quantum mechanics, are *locally inexplicable*. They cannot be explained, that is to say, without action at a distance.

To explain the 'inexplicable' we explain 'explicable'. For example the statistics of heart attacks in Lille and Lyons show strong correlations. The probability of M cases in Lyons and N in Lille, on a randomly chosen day, does not separate:

$$P(M, N) \neq P_1(M)P_2(N)$$

In fact when M is above average N also tends to be above average. You might shrug your shoulders and say 'coincidences happen all the time', or 'that's life'. Such an attitude is indeed sometimes advocated by otherwise serious people in the context of quantum philosophy. But outside that peculiar context, such an attitude would be dismissed as unscientific. The scientific attitude is that correlations cry out for explanation. And of course in the given example explanations are soon found. The weather is much the same in the two towns, and hot days are bad for heart attacks. The day of the week is exactly the same in the two towns, and Sundays are especially bad because of family quarrels and too much to eat. And so on. It seems reasonable to expect that if sufficiently many such causal factors can be identified and held fixed, the *residual* fluctuations will be independent, i.e.,

$$P(M, N | a, b, \lambda) = P_1(M | a, \lambda)P_2(N | b, \lambda) \tag{10}$$

where a and b are temperatures in Lyons and Lille respectively, λ denotes any number of other variables that might be relevant, and $P(M, N | a, b, \lambda)$ is the conditional probability of M cases in Lyons and N in Lille for *given* (a, b, λ). Note well that we already incorporate in (10) a hypothesis of 'local causality' or 'no action at a distance'. For we do not allow the first factor to depend on b, nor the second on a. That is, we do not admit the temperature in Lyons as a causal influence in Lille, and vice versa.

Let us suppose then that the correlations between A and B in the EPR experiment are likewise 'locally explicable'. That is to say we suppose that there are variables λ, which, if only we knew them, would allow decoupling of the fluctuations:

$$P(A, B | a, b, \lambda) = P_1(A | a, \lambda)P_2(B | b, \lambda) \tag{11}$$

We have to consider then some probability distribution $f(\lambda)$ over these complementary variables, and it is for the averaged probability

$$P(A, B | a, b) = \int d\lambda \, f(\lambda)P(A, B | a, b, \lambda) \tag{12}$$

that we have quantum mechanical predictions.

But not just any function $p(A, B|a, b)$ can be represented in the form (12). To see this it is useful to introduce the combination

$$E(a, b) = \begin{pmatrix} P(\text{yes}, \text{yes}|a, b) + P(\text{no}, \text{no}|a, b) \\ - P(\text{yes}, \text{no}|a, b) - P(\text{no}, \text{yes}|a, b) \end{pmatrix} \tag{13}$$

Then it is easy to show (Appendix 1) that if (12) holds, with however many variables λ and whatever distribution $\rho(\lambda)$, then follows the Clauser–Holt–Horne–Shimony[18] inequality

$$|E(a, b) + E(a, b') + E(a', b) - E(a', b')| \leqslant 2 \tag{14}$$

According to quantum mechanics, however, for example with some practical approximation to the EPRB gedanken set-up, we can have approximately (from (4))

$$E(a, b) = \left(\sin \frac{a - b}{2} \right)^2 - \left(\cos \frac{a - b}{2} \right)^2 = -\cos(a - b) \tag{15}$$

Taking for example

$$a = 0°, \quad a' = 90°, \quad b = 45°, \quad b' = -45° \tag{16}$$

We have from (15)

$$E(a, b) + E(a, b') + E(a', b) - E(a', b')$$
$$= -3\cos 45° + \cos 135° = -2\sqrt{2} \tag{17}$$

This is in contradiction with (14). Note that for such a contradiction it is not necessary to realize (15) accurately. A sufficiently close approximation is enough, for between (14) and (17) there is a factor of $\sqrt{2}$.

So the quantum correlations are locally inexplicable. To avoid the inequality we could allow P_1 in (11) to depend on b or P_2 to depend on a. That is to say we could admit the input at one end as a causal influence at the other end. For the set-up described this would be not only a mysterious long range influence – a non-locality or action at a distance in the loose sense – but one propagating faster than light (because $c\delta \ll L$) – a non-locality in the stricter and more indigestible sense.

It is notable that in this argument nothing is said about the locality, or even localizability, of the variable λ. These variables could well include, for example, quantum mechanical state vectors, which have no particular localization in ordinary space-time. It is assumed only that the outputs A and B, and the particular inputs a and b, are well localized.

5 Envoi

By way of conclusion I will comment on four possible positions that might be taken on this business – without pretending that they are the only possibilities.

First, and those of us who are inspired by Einstein would like this best, quantum mechanics may be *wrong* in sufficiently critical situations. Perhaps Nature is not so queer as quantum mechanics. But the experimental situation is not very encouraging from this point of view[19]. It is true that practical experiments fall far short of the ideal, because of counter inefficiencies, or analyzer inefficiencies, or geometrical imperfections, and so on. It is only with added assumptions, or conventional allowance for inefficiencies and extrapolation from the real to the ideal, that one can say the inequality is violated. Although there is an escape route there, it is hard for me to believe that quantum mechanics works so nicely for inefficient practical set-ups and is yet going to fail badly when sufficient refinements are made. Of more importance, in my opinion, is the complete absence of the vital *time* factor in existing experiments. The analyzers are not rotated during the flight of the particles. Even if one is obliged to admit some long range influence, it need not travel faster than light – and so would be much less indigestible. For me, then, it is of capital importance that Aspect[19,20] is engaged in an experiment in which the time factor is introduced.

Secondly, it may be that it is not permissible to regard the experimental settings a and b in the analyzers as independent variables, as we did[21]. We supposed them in particular to be independent of the supplementary variables λ, in that a and b could be changed without changing the probability distribution $\rho(\lambda)$. Now even if we have arranged that a and b are generated by apparently random radioactive devices, housed in separate boxes and thickly shielded, or by Swiss national lottery machines, or by elaborate computer programmes, or by apparently free willed experimental physicists, or by some combination of all of these, we cannot be *sure* that a and b are not significantly influenced by the same factors λ that influence A and B[21]. But this way of arranging quantum mechanical correlations would be even more mind boggling than one in which causal chains go faster than light. Apparently separate parts of the world would be deeply and conspiratorially entangled, and our apparent free will would be entangled with them.

Thirdly, it may be that we have to admit that causal influences *do* go faster than light. The role of Lorentz invariance in the completed theory would then be very problematic. An 'aether' would be the cheapest solution[22]. But the unobservability of this aether would be disturbing. So would the impossibility of 'messages' faster than light, which follows from ordinary relativistic quantum mechanics in so far as it is unambiguous and adequate for procedures we can actually perform. The exact elucidation of concepts like 'message' and 'we', would be a formidable challenge.

Fourthly and finally, it may be that Bohr's intuition was right – in that there is no reality below some 'classical' 'macroscopic' level. Then fundamental physical theory would remain fundamentally vague, until concepts like 'macroscopic' could be made sharper than they are today.

Appendix 1 – The position of Bohr

While imagining that I understand the position of Einstein[23,24], as regards the EPR correlations, I have very little understanding of the position of his principal opponent, Bohr. Yet most contemporary theorists have the impression that Bohr got the better of Einstein in the argument and are under the impression that they themselves share Bohr's views. As an indication of those views I quote a passage[25] from his reply to Einstein, Podolsky and Rosen. It is a passage which Bohr himself seems to have regarded as definitive, quoting it himself when summing up much later[26]. Einstein, Podolsky and Rosen had assumed that '... if, without in any way disturbing a system, we can predict with certainty the value of a physical quantity, then there exists an element of physical reality corresponding to this physical quantity'. Bohr replied: '... the wording of the above mentioned criterion... contains an ambiguity as regards the meaning of the expression "without in any way disturbing a system". Of course there is in a case like that just considered no question of a mechanical disturbance of the system under investigation during the last critical stage of the measuring procedure. But even at this stage there is essentially the question of *an influence on the very conditions which define the possible types of predictions regarding the future behaviour of the system*... their argumentation does not justify their conclusion that quantum mechanical description is essentially incomplete... This description may be characterized as a rational utilization of all possibilities of unambiguous interpretation of measurements, compatible with the finite and uncontrollable interaction between the objects and the measuring instruments in the field of quantum theory'.

Indeed I have very little idea what this means. I do not understand in what sense the word 'mechanical' is used, in characterizing the disturbances which Bohr does not contemplate, as distinct from those which he does. I do not know what the italicized passage means – 'an influence on the very conditions...'. Could it mean just that different experiments on the first system give different kinds of information about the second? But this was just one of the main points of EPR, who observed that one could learn *either* the position *or* the momentum of the second system. And then I do not understand the final reference to 'uncontrollable interactions between measuring instruments and objects', it seems just to ignore the essential point of EPR that in the absence of action at a distance, only the first system could be supposed disturbed by the first measurement and yet definite predictions become possible for the second system. Is Bohr just rejecting the premise – 'no action at a distance' – rather than refuting the argument?

Appendix 2 – Clauser–Holt–Horne–Shimony inequality

From (13) and (11)

$$E(a,b) = \int d\lambda f(\lambda) \{P_1(\text{yes}|a,\lambda) - P_1(\text{no}|a,\lambda)\}, \{P_2(\text{yes}|b,\lambda) - P_2(\text{no}|b,\lambda)\}$$

$$= \int d\lambda \, f(\lambda) \bar{A}(a,\lambda) \bar{B}(b,\lambda) \tag{18}$$

where $\bar{A}$ and $\bar{B}$ stand for the first and second curly brackets. Note that since the Ps are probabilities,

$$0 \leqslant P_1 \leqslant 1, \quad 0 \leqslant P_2 \leqslant 1$$

and it follows that

$$|\bar{A}(a,\lambda)| \leqslant 1, \quad |\bar{B}(b,\lambda)| \leqslant 1 \tag{19}$$

From (18)

$$E(a,b) \pm E(a,b') \leqslant \int d\lambda \, f(\lambda) \bar{A}(a,\lambda) [\bar{B}(b,\lambda) \pm \bar{B}(b',\lambda)]$$

so from (19)

$$|E(a,b) \pm E(a,b')| \leqslant \int d\lambda \, f(\lambda) |\bar{B}(b,\lambda) \pm \bar{B}(b',\lambda)|$$

likewise

$$|E(a',b) \mp E(a',b')| \leqslant \int d\lambda \, f(\lambda) |\bar{B}(b,\lambda) \mp \bar{B}(b',\lambda)|$$

Using again (19),

$$|\bar{B}(b,\lambda) \pm \bar{B}(b',\lambda)| + |\bar{B}(b,\lambda) \mp \bar{B}(b',\lambda)| \leqslant 2$$

and then from

$$\int d\lambda \, f(\lambda) = 1$$

follows

$$|E(a, b) \pm E(a, b')| + |E(a', b) \mp E(a', b')| \leqslant 2 \qquad (20)$$

which includes (14).

Notes and references

1 A. Einstein, B. Podolsky and N. Rosen, *Phys. Rev.* **46**, 777, (1935). For an introduction see the accompanying paper of F. Laloë.

2 D. Bohm, *Quantum Theory*. Englewood Cliffe, New Jersey (1951).

3 Note, however, that these *particular* phenomena were actually inferred from other quantum phenomena in advance of observation.

4 And perhaps romanticism. See P. Forman, Weimar culture, causality and quantum theory, 1918–1927, in *Historical Studies in the Physical Sciences*, vol. 3, 1–115. R. McCormach, ed. University of Pennsylvania Press, Philadelphia (1971).

5 M. Jammer, *The Philosophy of Quantum Mechanics*, John Wiley (1974), p. 204, quoting A. Petersen, *Bulletin of the Atomic Scientist* **19**, 12 (1963).

6 M. Jammer, *ibid*, p. 205, quoting W. Heisenberg, *Physics and Philosophy*, p. 160. Allen and Unwin, London (1958).

7 M. Jammer, *ibid*, p. 161, quoting E. Zilsel, P. Jordans Versuch, den Vitalismus quanten mechanisch zu retten, *Erkenntnis* **5**, (1935) 56–64.

8 The phrase is from a 1947 letter of Einstein to Born, Ref. 11, p. 158.

9 The accompanying paper of F. Laloë gives an introduction to quantum formalism.

10 And his followers. My own first paper on this subject (*Physics* **1**, 195 (1965).) starts with a summary of the EPR argument *from locality to* deterministic hidden variables. But the commentators have almost universally reported that it begins with deterministic hidden variables.

11 M. Born (editor), *The Born–Einstein-Letters*, p. 221. (Macmillan, London (1971).

12 M. Born, *ibid*, p. 176.

13 A. Einstein, *Dialectica*, 320, (1948). Included in a letter to Born, Ref. 11. p. 168.

14 B. d'Espagnat, *Scientific American*, p. 158. November 1979.

15 B. d'Espagnat, *A la Recherche du Réel*. Gauthier-Villars, Paris (1979).

16 'The number of young women is less than or equal to the number of woman smokers plus the number of young non-smokers.' (Ref. 15, p. 27). See also E. P. Wigner, *Am. J. Phys.* **38**, 1005 (1970).

17 Other discussions with some pretension to generality are: J. F. Clauser and M. A. Horne, *Phys. Rev.* **10D** (1974) 526; J. S. Bell, CERN preprint TH-2053 (1975), reproduced in *Epistemological Letters* (Association Ferd. Gonseth, CP 1081, CH-2051, Bienne) 9 (1976) 11; H. P. Stapp, *Foundations of Physics* **9** (1979) 1. Many other references are given in the reviews of Clauser and Shimony, and Pipkin in Ref. 19.

18 J. F. Clauser, R. A. Holt, M. A. Horne and A. Shimony, *Phys. Rev. Lett.* **23**, 880 (1969).

19 The experimental situation is surveyed in the accompanying paper of A. Aspect. See also: J. F. Clauser and A. Shimony, *Rep. Prog. Phys.* **41**, 1881 (1978); F. M. Pipkin, *Ann. Rev. Nuc. Sci.* (1978).

20 A. Aspect, *Phys. Rev.* **14D**, 1944 (1976).

21 For some explicit discussion of this, see contributions of Shimony, Horne, Clauser and Bell in *Epistemological Letters* (Association Ferdinand Gonseth, CP 1081, CH-2051, Bienne) **13**, p. 1 (1976); **15**, p. 79, (1977) and **18**, p. 1 (1978). See also Clauser and Shimony in Ref. 19.

22 P. H. Eberhard, *Nuovo Cimento* **46B**, 392 (1978).

23 But Max Jammer thinks that I misrepresent Einstein (Ref. 5, p. 254). I have defended my views in Ref. 24.

24 J. S. Bell, in *Frontier Problems in High Energy Physics, in honour of Gilberto Bernardini*. Scuola Normale, Pisa (1976).

25 N. Bohr, *Phys. Rev.* **48**, 696 (1935).

26 N. Bohr, in *Albert Einstein, Philosopher–Scientist*. P. A. Schilpp, Ed., Tudor, N.Y., (1949).

17

On the impossible pilot wave

1 Introduction

When I was a student I had much difficulty with quantum mechanics. It was comforting to find that even Einstein had had such difficulties for a long time. Indeed they had led him to the heretical conclusion that something was missing in the theory[1]: 'I am, in fact, rather firmly convinced that the essentially statistical character of contemporary quantum theory is solely to be ascribed to the fact that this (theory) operates with an incomplete description of physical systems.'

More explicitly,[2] in 'a complete physical description, the statistical quantum theory would... take an approximately analogous position to the statistical mechanics within the framework of classical mechanics...'.

Einstein did not seem to know that this possibility, of peaceful coexistence between quantum statistical predictions and a more complete theoretical description, had been disposed of with great rigour by J. von Neumann.[3] I myself did not know von Neumann's demonstration at first hand, for at that time it was available only in German, which I could not read. However I knew of it from the beautiful book by Born,[4] *Natural Philosophy of Cause and Chance*, which was in fact one of the highlights of my physics education. Discussing how physics might develop Born wrote: 'I expect... that we shall have to sacrifice some current ideas and to use still more abstract methods. However these are only opinions. A more concrete contribution to this question has been made by J. v. Neumann in his brilliant book, *Mathematische Grundlagen der Quantenmechanik*. He puts the theory on an axiomatic basis by deriving it from a few postulates of a very plausible and general character, about the properties of 'expectation values' (averages) and their representation by mathematical symbols. The result is that the formalism of quantum mechanics is uniquely determined by these axioms; in particular, no concealed parameters can be introduced with the help of which the indeterministic description could be transformed into a deterministic one. Hence if a future theory should be deterministic, it cannot be a modification of the present one but must be essentially different.

How this could be possible without sacrificing a whole treasure of well established results I leave to the determinists to worry about.'

Having read this, I relegated the question to the back of my mind and got on with more practical things.

But in 1952 I saw the impossible done. It was in papers by David Bohm.[5] Bohm showed explicitly how parameters could indeed be introduced, into nonrelativistic wave mechanics, with the help of which the indeterministic description could be transformed into a deterministic one. More importantly, in my opinion, the subjectivity of the orthodox version, the necessary reference to the 'observer,' could be eliminated.

Moreover, the essential idea was one that had been advanced already by de Broglie[6] in 1927, in his 'pilot wave' picture.

But why then had Born not told me of this 'pilot wave'? If only to point out what was wrong with it? Why did von Neumann not consider it? More extraordinarily, why did people go on producing 'impossibility' proofs,[7-12] after 1952, and as recently as 1978[13,14]? When even Pauli,[15] Rosenfeld,[16] and Heisenberg,[17] could produce no more devastating criticism of Bohm's version than to brand it as 'metaphysical' and 'ideological'? Why is the pilot wave picture ignored in text books? Should it not be taught, not as the only way, but as an antidote to the prevailing complacency? To show that vagueness, subjectivity, and indeterminism, are not forced on us by experimental facts, but by deliberate theoretical choice?

I will not attempt here to answer these questions. But, since the pilot wave picture still needs advertising, I will make here another modest attempt to publicize it, hoping that it may fall into the hands of a few of the many to whom even now it will be new. I will try to present the essential idea, which is trivially simple, so compactly, so lucidly, that even some of those who know they will dislike it may go on reading, rather than set the matter aside for another day.

2 A simple model

Consider a system whose wavefunction has one discrete argument, *a*, and one continuous argument, *x*, as well as time, *t*:

$$\Psi(a, x, t)$$
$$a = 1, 2, \ldots N$$
$$-\infty < x < +\infty$$

It might be a particle free to move in one-dimension and having an 'intrinsic spin.' Consider 'observables' O which involve only the spin, and so can be represented by finite matrices:

$$O\Psi(a, x) = \sum O(a, b)\Psi(b, x)$$

To 'measure' such an observable, suppose that we can contrive an interaction, with some external field, which is represented by the addition to the Hamiltonian of a term[3]

$$gO(\hbar/i)(\partial/\partial x)$$

where g is a coupling constant. Suppose for simplicity that the particle is infinitely massive, so that this interaction Hamiltonian is the complete Hamiltonian.[3] Then the Schrödinger equation is readily solved. It is convenient to introduce the eigenvectors of O

$$\alpha_n(a)$$

and corresponding eigenvalues

$$O_n$$

defined by

$$O\alpha_n(a) = O_n\alpha_n(a)$$

Then the initial state can be expanded

$$\Psi(a, x, o) = \sum_n \Phi_n(x)\alpha_n(a)$$

and the solution of the Schrödinger equation is

$$\Psi(a, x, t) = \sum_n \Phi_n(x - gO_nt)\alpha_n(a)$$

That is to say, the various wavepackets Φ move apart from one another, and after a sufficiently long time, whatever may have been the case initially, overlap very little. Then any probable result of a position measurement on the particle will correspond to a particular eigenvalue O_n, a particular O_n being obtained with probability given by the norm of the corresponding wavepacket Φ_n, i.e., by the strength of the corresponding eigenvector in the expansion of the initial state. We have here a model of something like a Stern–Gerlach experiment. Conventionally the process is said 'to measure observable O with result O_n.'

To complete this picture, *a la* de Broglie and Bohm, we add to the wavefunction Ψ a particle position

$$X(t)$$

If a position measurement is made at time t, then the result is $X(t)$, but even when no measurement is made $X(t)$ exists. The particle, in this picture, always has a definite position. The time evolution of particle position is determined by

$$(d/dt) X(t) = j(X(t), t)/\rho(X(t), t)$$

where

$$\rho(x, t) = \sum_a \Psi^*(a, x, t) \Psi(a, x, t)$$

$$j(x, t) = \sum_{a,b} \Psi^*(a, x, t) g O(a, b) \Psi(b, x, t)$$

Note that the Schrödinger equation implies the continuity equation

$$(\partial/\partial t)\rho + (\partial/\partial x)j = 0$$

It is assumed that, over many repetitions of the experiment, various $X(o)$ occur with the probability distribution

$$\rho(X(o), o) \, dX(o)$$

where ρ is given as above in terms of the initial wavefunction. Then it is a theorem that the probability distribution over $X(t)$ is

$$\rho(X(t), t) \, dX(t)$$

This is the conventional quantum distribution for position, and so we have the conventional predictions for the result of the Stern–Gerlach experiment. For the experiment, despite all the talk about 'spin,' is finally about position observations.

Note that in this theory probability enters once only, in connection with initial conditions, as in classical statistical mechanics. Thereafter the joint evolution of Ψ and X is perfectly deterministic.

Note that in this theory the wavefunction Ψ has the role of a physically real field, as real here as Maxwell's fields were for Maxwell. Quantum mechanics students sometimes have difficulty with the fact that in the pilot wave picture the particle position X and the argument of the wavefunction x are separate variables. But the situation, in this respect, is just that of Maxwell. He also had fields extending over space, and particles located at particular points. Of course the field at the particular point is that most immediately relevant for the motion of the particular particle.

Although Ψ is a real field it does not show up immediately in the result of a single 'measurement,' but only in the statistics of many such results. It is the de Broglie–Bohm variable X that shows up immediately each time. That X rather than Ψ is historically called a 'hidden' variable is a piece of historical silliness.

Note that from the present point of view the description of the experiment as 'measurement' of 'spin observable' O is an unfortunate one. Our particle has no internal degrees of freedom. It is guided however by a multicomponent field, and when this suffers the analogue of optical multiple refraction, the particle is dragged one way or another depending only on its initial position. We have here a very explicit illustration of the lesson taught by Bohr. Experimental results are products of the complete set-up, 'system' plus 'apparatus,' and should not be regarded as 'measurements' of preexisting properties of the 'system' alone.

3 The holes in the nets

It is easy to find good reasons for disliking the de Broglie–Bohm picture. Neither de Broglie[18] nor Bohm[19] liked it very much; for both of them it was only a point of departure. Einstein also[20] did not like it very much. He found it 'too cheap,' although, as Born[20] remarked, 'it was quite in line with his own ideas'.[21,22] But like it or lump it, it is perfectly conclusive as a counter example to the idea that vagueness, subjectivity, or indeterminism, are forced on us by the experimental facts covered by nonrelativistic quantum mechanics. What then is wrong with the impossibility proofs? Here I will consider only three of them, the most famous (incontestably), the most instructive (in my opinion), and the most recently published (to my knowledge). More, and more details, can be found elsewhere.[9,23-25]

It will be useful to denote by

$$R(O, \Psi(o), X(o))$$

the result of 'measuring' O in the above way, for given initial X and Ψ. This function can be calculated in principle by solving first the Schrödinger equation for Ψ and then solving the guiding equation for X. For some cases this has even been done explicitly.[26,27] Note well that the values taken by R are the eigenvalues of O.

The vital assumption in the famous proof of von Neumann is that, for operators connected by a linear relation,

$$O = pP + qQ$$

the results R are similarly related:

$$R(O, \Psi(o), X(o)) = pR(P, \Psi(o), X(o)) + qR(Q, \Psi(o), X(o))$$

Now this must certainly hold when averaged over $X(o)$ to give quantum expectation values. But it cannot possibly hold before averaging, for the individual results R are eigenvalues, and eigenvalues of linearly related operators are not linear related. For example let P and Q be components of spin angular momentum in perpendicular directions

$$P = S_x, \quad Q = S_y$$

and let O be the component along an intermediate direction

$$O = (P + Q)/\sqrt{2}$$

In the simple case of spin-1/2, the eigenvalues of O, P, Q, are all of magnitude 1/2, and the von Neumann requirement would read

$$\pm 1/2 = (\pm 1/2 \pm 1/2)/\sqrt{2}$$

– which is impossible indeed. Because the de Broglie–Bohm picture agrees with quantum mechanics in having the eigenvalues as the results of individual measurements – it is excluded by von Neumann. His 'very general and plausible' postulate is absurd.

More instructive is the Gleason–Jauch proof. I was told of it by J. M. Jauch in 1963. Not all of the powerful mathematical theorem of Gleason[28] is required, but only a corollary which is easily proved by itself.[9] (The idea was later rediscovered by Kochen and Specker[11]; see also Belinfante[24] and Fine and Teller[29].) Jauch saw that Gleason's theorem implied a result like that of von Neumann but with a weaker additivity assumption – for commuting operators only

$$[P, Q] = o$$

Since the eigenvalues of commuting operators are additive, additivity of the 'measurement' results is not manifestly absurd. Perhaps it seems particularly plausible when the commuting 'observables' involved are 'measured' at the same time. So let us go immediately to that case. It is sufficient to consider a complete set of orthogonal spin projection operators P_n, i.e., a set such that

$$P_n P_m = P_m P_n = P_n \delta_{nm}$$

and

$$\sum_n P_n = 1$$

The eigenvalues of such projection operators are all either zero or unity and, because the operators add to unity, the additivity hypothesis for 'measurement' results means simply that on 'measurement' one and only one of the operators will give unity, the others giving zero. It is easy to model this situation by an adaptation of the model described above. In the interaction Hamiltonian, gO is replaced by

$$\sum_n g_n P_n$$

The solution of the Schrödinger equation goes through as before in terms of simultaneous eigenvectors α of all the P_n. The various final wavepackets are displaced by distances g_n. The particle is found finally in one of these wavepackets; and, if the g_n are all different, this singles out one of the operators P_n as that for which the result of the 'measurement' is unity rather than zero. However the Gleason–Jauch argument depends also on another assumption. For a given operator P_1 it is possible (when the dimension N of the spin space exceeds 2) to find more than one set of other orthogonal projection operators to complete it:

$$1 = P_1 + P_2 + P_3 \ldots$$
$$= P_1 + P'_2 + P'_3 \ldots$$

where $P'_2 \ldots$ compute with P_1, and with one another, but not with $P_2 \ldots$. And the extra assumption is this: the result of 'measuring' P_1 is independent of which complementary set, $P_2 \ldots$ or $P'_2 \ldots$, is 'measured' at the same time. The de Broglie–Bohm picture does not respect this. Even though the two sets of operators have P_1 in common, the eigenvectors α are different, and the particle orbits $X(t)$ are different, as well as $\Psi(t)$, for given $X(o)$ and $\Psi(o)$. There is nothing unacceptable, or even surprising, about this. The Hamiltonians are different in the two cases. We are doing a different experiment when we arrange to 'measure' $P'_2 \ldots$ rather than $P_2 \ldots$ along with P_1. The apparent freedom of the Gleason–Jauch argument from implausible assumptions about incompatible 'observables' is illusory. In denying the Gleason–Jauch independence hypothesis, the de Broglie–Bohm picture illustrates rather the importance of the experimental set-up as a whole, as insisted on by Bohr. The Gleason–Jauch axiom is a denial of Bohr's insight.

The proof of Jost[13] concerns unstable 'identical' particles. He remarks that if decay time of similar nuclei were somehow determined in advance, by some parameters additional to the quantum wavefunction, then the nuclei would not be really identical and could not show the appropriate Fermi or Bose statistics. But again the difficulty disappears in the light of the pilot wave picture. The existing nonrelativistic version could not cope with beta decay. But it has no difficulty with alpha decay or fission (or even gamma decay[5]) when the unstable nuclei are regarded as composites of stable protons and neutrons. There is no problem in generalizing the de Broglie–Bohm picture to many particle systems.[5] The wavefunction is just that of ordinary quantum mechanics, and respects the usual symmetry or antisymmetry requirements. The added variables (in the simplest version of the theory[9,30,31]) are just particle positions, and the measured probability distributions of these will be those of quantum mechanics. Recognizing that it is always positions that we are in the end concerned with, all the statistical predictions of quantum mechanics are reproduced. This includes those phenomena associated with 'identity of particles'.[5] The anticipated difficulty does not arise.

3 Morals

The first moral of this story is just a practical one. Always test your general reasoning against simple models.

The second moral is that in physics the only observations we must consider are position observations, if only the positions of instrument pointers. It is a great merit of the de Broglie–Bohm picture to force us to consider this fact. If you make axioms, rather than definitions and theorems, about the 'measurement' of anything else, then you commit redundancy and risk inconsistency.

A final moral concerns terminology. Why did such serious people take so seriously axioms which now seem so arbitrary? I suspect that they were misled by the pernicious misuse of the word 'measurement' in contemporary theory. This word very strongly suggests the ascertaining of some preexisting property of some thing, any instrument involved playing a purely passive role. Quantum experiments are just not like that, as we learned especially from Bohr. The results have to be regarded as the joint product of 'system' and 'apparatus,' the complete experimental set-up. But the misuse of the word 'measurement' makes it easy to forget this and then to expect that the 'results of measurements' should obey some simple logic in which the apparatus is not mentioned. The resulting difficulties soon show that any such logic is not ordinary logic. It is my impression that the whole vast subject of 'Quantum Logic' has arisen in this way from the misuse of a word. I am convinced that the word 'measurement' has now been so abused that the field would be significantly advanced by banning its use altogether, in favour for example of the word 'experiment.'

There are surely other morals to be drawn here, if not by physicists then by historians and sociologists.[32,33]

Of the various impossibility proofs, only those concerned with local causality[34-37] seem now to retain some significance outside special formalisms. The de Broglie–Bohm theory is not a counter example in this case. Indeed it was the explicit representation of quantum nonlocality in that picture which started a new wave of investigation in this area. Let us hope that these analyses also may one day be illuminated, perhaps harshly, by some simple constructive model.

However that may be, long may Louis de Broglie continue to inspire those who suspect that what is proved by impossibility proofs is lack of imagination.

Acknowledgements

I have profited from comments by M. Bell, E. Etim, K. V. Laurikainen, J. M. Leinaas, and J. Kupsch.

Note added in proof

I am sorry to have missed, before writing the above, an early paper by E. Specker (*Dialectia* **14**, 239 (1960), or in C. A. Hooker, ed., *The Logico-Algebraic Approach to Quantum Mechanics*, p. 135. Reidel, Dordrecht, (1975)). It announced already what I have called the Gleason–Jauch result. Specker did not know of the work of Gleason, but mentioned rather the possibility of an 'elementary geometrical argument' – presumably of the kind that I myself gave later[9] as a preliminary to criticism of the axioms.

References

1 P. A. Schilpp, Ed., *Albert Einstein, Philosopher Scientist*. p. 666. Tudor, New York (1949).
2 P. A. Schilpp, Ed., *Albert Einstein, Philosopher Scientist*. p. 672. Tudor, New York (1949).
3 J. von Neumann, *Mathematisch Grundlagen der Quantenmechanik*. Springer Verlag, Berlin (1932); English translation: Princeton University Press (1955).
4 M. Born, *Natural Philosophy of Cause and Chance*. Clarendon, Oxford (1949).
5 D. Bohm, *Phys. Rev.* **85**, 165, 180 (1952).
6 L. de Broglie, in *Rapport au V'ieme Congres de Physique Solvay*. Gauthier-Villars, Paris (1930).
7 J. M. Jauch and C. Piron, *Helvetica Physica Acta* **36**, 827 (1963); *Rev. Mod. Phys.* **40**, 228 (1966).
8 J. M. Jauch, private communication (1963).
9 J. S. Bell, SLAC-PUB-44, Aug. 1964; *Rev. Mod. Phys.* **38**, 447 (1966).
10 B. Misra, *Nuovo Cimento* **47**, 843 (1967).
11 S. Kochen and E. P. Specker, *J. Math. Mech.* **17**, 59 (1967).
12 S. P. Gudder, *Rev. Mod. Phys.* **40**, 229 (1968); *J. Math. Phys.* **9**, 1411 (1968).
13 R. Jost, in *Some Strangeness in the Proportion*. p. 252. Addison-Wesley, Reading (1980).

14 H. Woolf, ed., *Some Strangeness in the Proportion*. Addison-Wesley, Reading (1980).

15 W. Pauli, in A. George, ed., *Louis de Broglie, Physicien et Penseur*. Albin Michel, Paris (1953).

16 L. Rosenfeld, in A. George, ed., *Louis de Broglie, Physicien et Penseur*. p. 43. Albin Michel, Paris (1953).

17 W. Heisenberg, in W. Pauli, ed., *Niels Bohr and the Development of Physics*. Pergamon, London (1955).

18 L. de Broglie, *Found. Phys.* **1**, 5 (1970).

19 D. Bohm, *Wholeness and the Implicate Order*, Routledge and Kegan Paul, London (1980).

20 M. Born, ed., *The Born–Einstein Letters*, p. 192, and letters 81, 84, 86, 88, 97, 99, 103, 106, 108, 110, 115, and 116. Macmillan, London (1971).

21 E. P. Wigner, in *Some Strangeness in the Proportion*. p. 463. Addison-Wesley, Reading (1980).

22 J. S. Bell, in *Proc. Symposium on Frontier Problems in High Energy Physics*, in honor of Gilberto Bernardini on his 70th birthday (Scuola Normale Superiore, Pisa, 1976).

23 M. Mugur-Schächter, *Étude du Charactère Complet de la Théorie Quantique*. Gauthier-Villars, Paris (1964).

24 F. J. Belinfante, *A Survey of Hidden Variable-Theories*. Pergamon, London (1973).

25 M. Jammer, *The Philosophy of Quantum Mechanics*. Wiley, New York (1974).

26 C. Phillipidas, C. Dewdney, and B. J. Hiley, *Nuov. Cim.* **52B**, 15 (1979).

27 C. Dewdney and B. J. Hiley, *Found. Phys.* **12**, 27 (1982).

28 A. M. Gleason, *J. Math. Mech.* **6**, 885 (1957).

29 A. Fine and P. Teller, *Found. Phys.* **8**, 629 (1978).

30 J. S. Bell, *Inter. J. of Quantum Chem., Quantum Chemistry Symposium No. 14* (Wiley, New York, 1980).

31 C. J. Isham, R. Penrose, and D. W. Sciama, eds., *Quantum Gravity 2* p. 611. Clarendon, Oxford (1980).

32 P. Forman, Weimar Culture, Causality, and Quantum Theory 1918–1927, in R. McCormach, ed., *Historical Studies in Physical Sciences 3*, pp. 1–115. Univ. of Pennsylvania Press, Philadelphia (1971).

33 T. J. Pinch, 'What Does a Proof Do if it Does Not Prove?, A Study of the Social Conditions and Metaphysical Divisions Leading to David Bohm and John von Neumann Failing to Communicate in Quantum Physics,' in E. Mendelsohn, P. Weingart, and R. Whitly, eds., *The Social Production of Scientific Knowledge* pp. 171–215. Reidel, Dordrecht (1977).

34 J. Clauser and A. Shimony, *Rep. Prog. Phys.* **41**, 1881 (1978).

35 F. M. Pipkin, *Advances in Atomic and Molecular Physics 14*. p. 281. Academic Press, New York (1979).

36 F. Selleri and G. Tarozzi, *Riv. Nuov. Cim.* **4** (2) (1981).

37 B. d'Espagnat, *A la Recherche du Réel*. Gauthier Villars, Paris (1979).

18

Speakable and unspeakable in quantum mechanics

'....the history of cosmic theories may without exaggeration be called a history of collective obsessions and controlled schizophrenias; and the manner in which some of the most important individual discoveries were arrived at reminds one of a sleepwalker's performance....'

This is a quotation from A. Koestler's book *The Sleepwalkers*. It is an account of the Copernican revolution, with Copernicus, Kepler, and Galilei as heroes. Koestler was of course impressed by the magnitude of the step made by these men. He was also fascinated by the manner in which they made it. He saw them as motivated by irrational prejudice, obstinately adhered to, making mistakes which they did not discover, which somehow cancelled at the important points, and unable to recognize what was important in their results, among the mass of details. He concluded that they were not really aware of what they were doing...sleepwalkers. I thought it would be interesting to keep Koestler's thesis in mind as we hear at this meeting about contemporary theories from contemporary theorists.

For many decades now our fundamental theories have rested on the two great pillars to which this meeting is dedicated: quantum theory and relativity. We will see that the lines of research opened up by these theories remain splendidly vital. We will see that order is brought into a vast and expanding array of experimental data. We will see even a continuing ability to get ahead of the experimental data... as with the existence and masses of the W and Z mesons. Perhaps this more than anything convinces us that there is truth in what is done.

In the manner in which this progress is made, will we see again any elements of Koestler's picture? Certainly we will see nothing like the obsessive commitment of the old heroes to their hypotheses. Our theorists take up and put down hypotheses with light hearts, playfully. There is no religious intensity in it. And certainly no fear of becoming involved in litigation with the religious authorities. As for technical mistakes, our theorists do not make them. And they see at once what is important and what is detail. So it is another feature of contemporary progress which reminds me of the title of Koestler's book. This progress is made in spite of the fundamental obscurity in quantum mechanics. Our theorists stride through that obscurity unimpeded... sleepwalking?

The progress so made is immensely impressive. If it is made by sleep-walkers, is it wise to shout 'wake up'? I am not sure that it is. So I speak now in a very low voice.

In a moment I will try to locate the 'Problem' of quantum mechanics. But first let me argue against a myth... that quantum theory had undone somehow the Copernican revolution. From those who made that revolution we learned that the world is more intelligible when we do not imagine ourselves to be at the centre of it. Does not quantum theory again place 'observers'...us...at the centre of the picture? There is indeed much talk of 'observables' in quantum theory books. And from some popular presentations the general public could get the impression that the very existence of the cosmos depends on our being here to observe the observ-ables. I do not know that this is wrong. I am inclined to hope that we are indeed that important. But I see no evidence that it is so in the success of contemporary quantum theory.

So I think it is not right to tell the public that a central role for conscious mind is integrated into modern atomic physics. Or that 'information' is the real stuff of physical theory. It seems to me irresponsible to suggest that technical features of contemporary theory were anticipated by the saints of ancient religions... by introspection.

The only 'observer' which is essential in orthodox practical quantum theory is the inanimate apparatus which amplifies microscopic events to macroscopic consequences. Of course this apparatus, in laboratory experiments, is chosen and adjusted by the experimenters. In this sense the outcomes of experiments are indeed dependent on the mental processes of the experimenters! But once the apparatus is in place, and functioning untouched, it is a matter of complete indifference... according to ordinary quantum mechanics... whether the experimenters stay around to watch, or delegate such 'observing' to computers.

Why this necessity to refer to 'apparatus' when we would discuss quantum phenomena? The physicists who first came upon such phenomena found them so bizarre that they despaired of describing them in terms of ordinary concepts like space and time, position and velocity. The founding fathers of quantum theory decided even that no concepts could possibly be found which could permit direct description of the quantum world. So the theory which they established aimed only to describe systematically the response of the apparatus. And what more, after all, is needed for application? It is as if our friends could not find words to tell us about the very strange places where they went on holiday. We could see for ourselves whether they came back browner or fatter. This would be enough for us to be able to advise other friends, who might wish to be browner or fatter, about those strange places. Our apparatus visits the microscopic world for us, and we see what happens to it as a result.

The 'Problem' then is this: how exactly is the world to be divided into speakable apparatus...that we can talk about...and unspeakable quantum system that we can not talk about? How many electrons, or atoms, or molecules, make an 'apparatus'? The mathematics of the ordinary theory requires such a division, but says nothing about how it is to be made. In practice the question is resolved by pragmatic recipes which have stood the test of time, applied with discretion and good taste born of experience. But should not fundamental theory permit exact mathematical formulation?

Now in my opinion the founding fathers were in fact wrong on this point. The quantum phenomena do *not* exclude a uniform description of micro and macro worlds...system and apparatus. It is *not* essential to introduce a vague division of the world of this kind. This was indicated already by de Broglie in 1926, when he answered the conundrum

wave or particle?

by

wave *and* particle.

But by the time this was fully clarified by Bohm in 1952, few theoretical physicists wanted to hear about it. The orthodox line seemed fully justified by practical success. Even now the de Broglie–Bohm picture is generally ignored, and not taught to students. I think this is a great loss. For that picture exercises the mind in a very salutary way.

The de Broglie–Bohm picture disposes of the necessity to divide the world somehow into system and apparatus. But another problem is brought into focus. This picture, and indeed, I think, any sharp formulation of quantum mechanics, has a very surprising feature: the consequences of events at one place propagate to other places faster than light. This happens in a way that we cannot use for signalling. Nevertheless it is a gross violation of relativistic causality. Moreover the specific quantum phenomena that require such superluminal explanation have been largely realised in the laboratory...especially by Aspect, Dalibard, and Roger, in Paris in 1982 (*Phys. Rev. Lett.* **49**, 1804 (1982)).

For me then this is the real problem with quantum theory: the apparently essential conflict between any sharp formulation and fundamental relativity. That is to say, we have an apparent incompatibility, at the deepest level, between the two fundamental pillars of contemporary theory... and of our meeting. I am glad therefore that in some of the sessions we will stand back from the impressive technical details of current progress to review this strange situation. It may be that a real synthesis of quantum and relativity theories requires not just technical developments but radical conceptual renewal.

19

Beables for quantum field theory

Dedicated to Professor D. Bohm

1 Introduction

Bohm's 1952 papers[1, 2] on quantum mechanics were for me a revelation. The elimination of indeterminism was very striking. But more important, it seemed to me, was the elimination of any need for a vague division of the world into 'system' on the one hand, and 'apparatus' or 'observer' on the other. I have always felt since that people who have not grasped the ideas of those papers...and unfortunately they remain the majority...are handicapped in any discussion of the meaning of quantum mechanics.

When the cogency of Bohm's reasoning is admitted, a final protest is often this: it is all nonrelativistic. This is to ignore that Bohm himself, in an appendix to one of the 1952 papers[2], already applied his scheme to the electromagnetic field. And application to scalar fields is straightforward[3]. However until recently[4,5], to my knowledge, no extension covering Fermi fields had been made. Such an extension will be sketched here. The need for Fermi fields might be questioned. Fermions might be composite structures of some kind[6]. But also they might not be, or not all. The present exercise will not only include Fermi fields, but even give them a central role. The dependence on the ideas of de Broglie[7] and Bohm[1,2], and also on my own simplified extension to cover spin[8,9,10], will be manifest to those familiar with these things. However no such familiarity will be assumed.

A preliminary account of these notions[5] was entitled 'Quantum field theory without observers, or observables, or measurements, or systems, or apparatus, or wavefunction collapse, or anything like that'. This could suggest to some that the issue in question is a philosophical one. But I insist that my concern is strictly professional. I think that conventional formulations of quantum theory, and of quantum field theory in particular, are unprofessionally vague and ambiguous. Professional theoretical physicists ought to be able to do better. Bohm has shown us a way.

It will be seen that all the essential results of ordinary quantum field theory are recovered. But it will be seen also that the very sharpness of the reformulation brings into focus some awkward questions. The construction of the scheme is not at all unique. And Lorentz invariance plays a strange, perhaps incredible role.

2 Local beables

The usual approach, centred on the notion of 'observable', divides the world somehow into parts: 'system' and 'apparatus'. The 'apparatus' interacts from time to time with the 'system', 'measuring' 'observables'. During 'measurement' the linear Schrödinger evolution is suspended, and an ill-defined 'wavefunction collapse' takes over. There is nothing in the mathematics to tell what is 'system' and what is 'apparatus', nothing to tell which natural processes have the special status of 'measurements'. Discretion and good taste, born of experience, allow us to use quantum theory with marvelous success, despite the ambiguity of the concepts named above in quotation marks. But it seems clear that in a serious fundamental formulation such concepts must be excluded.

In particular we will exclude the notion of 'observable' in favour of that of '*be*able'. The beables of the theory are those elements which might correspond to elements of reality, to things which exist. Their existence does not depend on 'observation'. Indeed observation and observers must be made out of beables.

I use the term 'beable' rather than some more committed term like 'being'[11] or 'beer'[12] to recall the essentially tentative nature of any physical theory. Such a theory is at best a *candidate* for the description of nature. Terms like 'being', 'beer', 'existent'[11,13], etc., would seem to me lacking in humility. In fact 'beable' is short for 'maybe-able'.

Let us try to promote some of the usual 'observables' to the status of beables. Consider the conventional axiom:

the probability of observables $(A, B, \ldots)$ (1)
if observed at time t
being observed to be $(a, b, \ldots)$
is

$$\sum_q |\langle a, b, \ldots q | t \rangle|^2$$

where q denotes additional quantum numbers which
together with the eigenvalues $(a, b, \ldots)$
form a complete set.

This we replace by

> the probability of beables $(A, B, \ldots)$ (2)
> at time t
> being $(a, b, \ldots)$
> is

$$\sum_q |\langle a, b, \ldots q | t \rangle|^2$$

where q denotes additional quantum numbers which together with the eigenvalues $(a, b, \ldots)$ form a complete set.

Not all 'observables' can be given beable status, for they do not all have simultaneous eigenvalues, i.e. do not all commute. It is important to realize therefore that most of these 'observables' are entirely redundant. What is essential is to be able to define the positions of things, including the positions of instrument pointers or (the modern equivalent) of ink on computer output.

In making precise the notion 'positions of things' the energy density $T_{00}(x)$ comes immediately to mind. However the commutator

$$[T_{00}(x), T_{00}(y)]$$

is not zero, but proportional to derivatives of delta functions. So the $T_{00}(x)$ do not have simultaneous eigenvalues for all x. We would have to devise some new way of specifying a joint probability distribution.

We fall back then on a second choice – fermion number density. The distribution of fermion number in the world certainly includes the positions of instruments, instrument pointers, ink on paper, ... and much much more.

For simplicity we replace the three-space continuum by a dense lattice, keeping time t continuous (and real!). Let the lattice points be enumerated by

$$l = 1, 2, \ldots L$$

where L is very large. Define lattice point fermion number operators

$$\psi^+(l)\psi(l)$$

where summation over Dirac indices and over all Dirac fields is understood. The corresponding eigenvalues are integers

$$F(l) = 1, 2, \ldots 4N$$

where N is the number of Dirac fields. The fermion number configuration of the world is a list of such integers

$$n = (F(1), F(2), \dots F(L))$$

We suppose the world to have a definite such configuration at every time t:

$$n(t)$$

The lattice fermion number are the local beables of the theory, being associated with definite positions in space. The state vector $|t\rangle$ also we consider as a beable, although not a local one. The complete specification of our world at time t is then a combination

$$(|t\rangle, n(t)) \tag{3}$$

It remains to specify the time evolution of such a combination.

3 Dynamics

For the time evolution of the state vector we retain the ordinary Schrödinger equation,

$$\mathrm{d}/\mathrm{d}t|t\rangle = -\mathrm{i}H|t\rangle \tag{4}$$

where H is the ordinary Hamiltonian operator.

For the fermion number configuration we prescribe a stochastic development. In a small time interval $\mathrm{d}t$ configuration m jumps to configuration n with transition probability

$$\mathrm{d}t\, T_{nm} \tag{5}$$

where

$$T_{nm} = J_{nm}/D_m \tag{6}$$

$$J_{nm} = \sum_{qp} 2\,\mathrm{Re}\langle t|nq\rangle\langle nq| -\mathrm{i}H|mp\rangle\langle mp|t\rangle \tag{7}$$

$$D_m = \sum_q |\langle mq|t\rangle|^2 \tag{8}$$

provided $J_{nm} > 0$, but

$$T_{nm} = 0 \quad \text{if} \quad J_{nm} \leqslant 0 \tag{9}$$

From (5) the evolution of a probability distribution P_n over configurations n is given by

$$\mathrm{d}/\mathrm{d}t\, P_n = \sum_m (T_{nm}P_m - T_{mn}P_n) \tag{10}$$

Compare this with a mathematical consequence of the Schrödinger equation (4):

$$d/dt|\langle nq|t\rangle|^2 = \sum_{mp} 2\,\mathrm{Re}\langle t|nq\rangle\langle nq|-iH|mp\rangle\langle mp|t\rangle$$

or

$$d/dtD_n = \sum_m J_{nm} = \sum_m (T_{nm}D_m - T_{mn}D_n) \tag{11}$$

If we assume that at some initial time

$$P_n(0) = D_n(0) \tag{12}$$

then from (11) the solution of (10) is

$$P_n(t) = D_n(t) \tag{13}$$

Envisage then the following situation. In the beginning God chose 3-space and 1-time, a Hamiltonian H, and a state vector $|0\rangle$. Then She chose a fermion configuration $n(0)$. This She chose at random from an ensemble of possibilities with distribution $D(0)$ related to the already chosen state vector $|0\rangle$. Then She left the world alone to evolve according to (4) and (5).

It is notable that although the probability distribution P in (13) is governed by D and so by $|t\rangle$, the latter is not to be thought of as just a way of expressing the probability distribution. For us $|t\rangle$ is an independent beable of the theory. Otherwise its appearance in the transition probabilities (5) would be quite unintelligible.

The stochastic transition probabilities (5) replace here the deterministic guiding equation of the de Broglie–Bohm 'pilot wave' theory. The introduction of a stochastic element, for beables with discrete spectra, is unwelcome, for the reversibility[14] of the Schrödinger equation strongly suggests that quantum mechanics is not fundamentally stochastic in nature. However I suspect that the stochastic element introduced here goes away in some sense in the continuum limit.

4 OQFT and BQFT

OQFT is 'ordinary' 'orthodox' 'observable' quantum field theory, whatever that may mean. BQFT is de Broglie–Bohm beable quantum field theory. To what extent do they agree? The main difficulty with this question is the absence of any sharp formulation of OQFT. We will consider two different ways of reducing the ambiguity.

In OQFT1 the world is considered as one big experiment. God prepared it at the initial time $t = 0$, and let it run. At some much later time T She will return to judge the outcome. In particular She will observe the contents of all the physics journals. This will include of course the records of our own little experiments – as distributions of ink on paper, and so of fermion number. From (13) the OQFT1 probability D that God will observe one configuration rather than another is identical with the BQFT probability P that the configuration *is* then one thing rather than another. In this sense there is complete agreement between OQFT1 and BQFT on the result of God's big experiment – including the results of our little ones.

OQFT1, in contrast with BQFT, says nothing about events in the system in between preparation and observation. However adequate this may be from an Olympian point of view, it is rather unsatisfactory for us. We live in between creation and last judgement – and imagine that we experience events. In this respect another version of OQFT is more appealing. In OQFT2, whenever the state can be resolved into a sum of two (or more) terms

$$|t\rangle = |t, 1\rangle + |t, 2\rangle \tag{14}$$

which are 'macroscopically different', then in disregard for the Schrödinger equation the state 'collapses' somehow into one term or the other:

$$|t\rangle \rightarrow N_1^{-1/2}|t, 1\rangle \quad \text{with probability } N_1 \tag{15}$$
$$|t\rangle \rightarrow N_2^{-1/2}|t, 2\rangle \quad \text{with probability } N_2$$

where

$$N_1 = |\langle t, 1|t, 1\rangle| \quad N_2 = |\langle t, 2|t, 2\rangle| \tag{16}$$

In this way the state is always, or nearly always, macroscopically unambiguous and defines a macroscopically definite history for the world. The words 'macroscopic' and 'collapse' and terribly vague. Nevertheless this version of OQFT is probably the nearest approach to a rational formulation of how we use quantum theory in practice.

Will OQFT2 agree with OQFT1 and BQFT at the final time T? This is the main issue in what is usually called 'the Quantum Measurement Problem'. Many authors, analyzing many models, have convinced themselves that the state vector collapse of OQFT2 is consistent with the Schrödinger equation of OQFT1 'for all practical purposes'[15]. The idea is that even when we retain both components in (13), evolving as required by the Schrödinger equation, they remain so different as not to interfere in the calculation of anything of interest. The following sharper form of this hypothesis seems plausible to me: the macroscopically distinct components remain so different, for a very long time, as not to interfere in the calculation of D and $J^{(5)}$. In so far as this is true, the trajectories of OQFT2 and BQFT will agree macroscopically.

5 Concluding remarks

We have seen that BQFT is in complete accord with OQFT1 as regards the final outcome. It is plausibly consistent with OQFT2 in so far as the latter is unambiguous. BQFT has the advantage over OQFT1 of being relevant at all times, and not just at the final time. It is superior to OQFT2 in being completely formulated in terms of unambiguous equations.

Yet even BQFT does not inspire complete happiness. For one thing there is nothing unique about the choice of fermion number density as basic local beable. We could have others instead, or in addition. For example the Higg's fields of contemporary gauge theories could serve very well to define 'the positions of things'. Other possibilities have been considered by K. Baumann[4]. I do not see how this choice can be made experimentally significant, so long as the final results of experiments are defined so grossly as by the positions of instrument pointers, or of ink on paper.

And the status of Lorentz invariance is very curious. BQFT agrees with OQFT on the result of the Michelson–Morley experiment, and so on. But the formulation of BQFT relies heavily on a particular division of space-time into space and time. Could this be avoided?

There is indeed a trivial way of imposing Lorentz invariance[4]. We can imagine the world to differ from vacuum over only a limited region of infinite Euclidean space (we forget general relativity here). Then an overall centre of mass system is defined. We can simply assert that our equations hold in this centre of mass system. Our scheme is then Lorentz invariant. Many others could be made Lorentz invariant in the same way...for example Newtonian mechanics. But such Lorentz invariance would not imply a null result for the Michelsen–Morley experiment... which could detect motion relative to the cosmic mass centre. To be predictive, Lorentz invariance must be supplemented by some kind of locality, or separability, consideration. Only then, in the case of a more or less isolated object, can motion relative to the world as a whole be deemed more or less irrelevant.

I do not know of a good general formulation of such a locality requirement. In classical field theory, part of the requirement could be formulation in terms of differential (as distinct from integral) equations in $3 + 1$ dimensional space-time. But it seems clear that quantum mechanics requires a much bigger configuration space. One can formulate a locality requirement by permitting arbitrary external fields, and requiring that variation thereof have consequences only in their future light cones. In that case the fields could be used to set measuring instruments, and one comes into difficulty with quantum predictions for correlations related to those of Einstein, Podolsky, and Rosen[18]. But the introduction of external fields is questionable. So I am unable to prove, or even formulate clearly, the proposition that a sharp formulation of quantum field theory, such as that set out here, must disrespect serious Lorentz invariance. But it seems to me that this is probably so.

As with relativity before Einstein, there is then a preferred frame in the formulation of the theory... but it is experimentally indistinguishable[20,21,22]. It seems an eccentric way to make a world.

Notes and references

1 D. Bohm, *Phys. Rev.* **85**, 166 (1952).
2 D. Bohm, *Phys. Rev.* **85**, 180 (1952).
3 D. Bohm and B. Hiley, *Found. Phys.* **14**, 270 (1984).
4 K. Baumann, preprint, Graz (1984).
5 J. S. Bell, *Phys. Rep.* **137**, 49–54 (1986).
6 T. H. R. Skyrme, *Proc. Roy. Soc.* **A260**, 127 (1961). A. S. Goldhaber, *Phys. Rev. Lett.* **36**, 1122 (1976). F. Wilczek and A. Zee, *Phys. Rev. Lett.* **51**, 2250 (1983).
7 L. de Broglie, *Tentative d'Interpretation Causale et Nonlineaire de la Mechanique Ondulatoire.* Gauthier–Villars, Paris (1956).
8 J. S. Bell, *Rev. Mod. Phys.* **38**, 447 (1966).
9 J. S. Bell, in *Quantum Gravity*, p. 611, Edited by Isham, Penrose and Sciama, Oxford (1981), (originally TH. 1424-CERN, 1971 Oct 27).
10 J. S. Bell, *Found. Phys.* **12**, 989 (1982).
11 A. Shimony, *Epistemological Letters*, Jan 1978, 1.
12 B. Zumino, private communication.
13 B. d'Espagnat, *Phys. Rep.*, **110**, 202–63. (1984).
14 I ignore here the small violation of time reversibility that has shown up in elementary particle physics. It could be of 'spontaneous' origin. Moreover PCT remains good.
15 This is touched on in Refs. 9, 16, and 17, and in many papers in the anthology of Wheeler and Zurek Ref.19.
16 J. S. Bell, *Helv. Phys. Acta* **48**, 93. (1975).
17 J. S. Bell, *Int. J. Quant. Chem.*: Quantum Chemistry Symposium **14**, p. 155. (1980).
18 J. S. Bell, *Journal de Physique*, Colloque C2, suppl. au no. 3, Tome 42, p. C2–41, mars 1981.
19 J. A. Wheeler and W. H. Zurek (editors), *Quantum Theory and Measurement.* Princeton University Press, Princeton 1983.
20 J. S. Bell, in *Determinism, Causality, and Particles*, p. 17. Edited by M. Flato *et al.* Dordrecht-Holland, D. Reidel (1976).
21 P. H. Eberhard, *Nuovo Cimento* **46B**, 392 (1978).
22 K. Popper, *Found. Phys.* **12**, 971 (1982).

20

Six possible worlds of quantum mechanics

I suppose one could imagine laws of physics which would dictate that a world be exactly so, and not otherwise, allowing no detail to be varied. But what could dictate that those laws of physics be 'the' laws of physics? By considering a spectrum of possible laws, one could again consider a spectrum of possible worlds.

In fact the laws of physics of our actual world, as presently understood, have no such dictatorial character. So that even with the laws given, a spectrum of different worlds is possible. There are two kinds of freedom. Although the laws say something about how a given state of the world may develop, they say nothing (or anyway very little) about in what state the world should start. So, to begin with, we have freedom as regards 'initial conditions'. To go on with, the future that can evolve from a given present is not uniquely determined, according to contemporary orthodoxy. The laws list various possibilities, and attach to them various probabilities.

The relation between the set of possibilities and the unique actuality which emerges is quite peculiar in modern 'quantum theory' – the contemporary all-embracing basic physical theory. The absence of determinism,

Fig. 1. Electron gun.

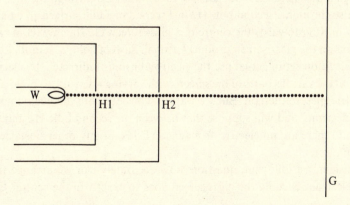

the probabilistic nature of the assertions of the theory, is already a little peculiar... at least in the light of pre-twentieth-century 'classical' physics. But after all everyday life, if not classical physics, prepares us very well for the idea that not everything is predictable, that chance is important. So it is not in the indeterminism that the real surprise of quantum theory lies. There are other aspects of quantum theory for which neither classical physics nor everyday life prepares us at all.

As a result some very different conceptions, and some very strange ones, have arisen, about how the visible phenomena might be incorporated into a coherent theoretical picture. It is to several such very different possible worlds that the title of this essay refers, rather than the permissible variation of incidental detail within each. Before giving some account of these schemes, we recall some of the phenomena with which they have to cope.

Atoms of matter can be pictured, to some extent, as small solar systems. The electrons circulate about the nucleus as do the planets about the sun. Since Newton we have very accurate laws for the motion of planets about suns, and since Einstein laws more accurate still. Attempts to apply similar laws to electrons in atoms meet with conspicuous failure. It was such failure that led to the development of 'quantum' mechanics to replace 'classical' mechanics. Of course our ideas about electrons in atoms are arrived at only indirectly, from the behaviour of pieces of matter containing many electrons in many atoms. But in extreme conditions quantum ideas are essential even for 'free' electrons, extracted from atoms, such as those which create the image on a television screen. It is in this simpler context that we will introduce the quantum ideas here.

In the 'electron gun' of a television set (Fig. 1) a wire W is heated, by passage of an electric current, so that some electrons 'boil off'. These are attracted to a metal surface, by an electric field, and some of them pass through a hole in it, H1. And some of those that pass through the hole H1 pass also through a second hole H2 in a second metallic surface, to emerge finally moving towards the centre of a glass screen G. The impact of each electron on the glass screen produces a small flash of light, a 'scintillation'. In a television set in actual use the electron beam is redirected, by electric fields, to the various parts of the screen, with varying intensity, to build up a complete picture thereon. But we want to consider here the behaviour of 'free' electrons, and will suppose that between the second hole H2 and the screen G there are no electric or magnetic fields, or any other obstacle to 'free' motion.

Consider the following question: how accurately can we arrange that each electron reaching the glass screen does so exactly in the centre? One thing to avoid, to this end, is that different electrons jostle one another. This can be done by 'pulsing' (i.e. by applying for only a very short time) the electric field that attracts electrons from W towards H1, and by making H1 very small. Then it becomes very unlikely that more than one electron will emerge from the hole H1 on a given occasion. Then one might reasonably think that to avoid any particle striking the glass screen off centre it is sufficient to make H2 as well as H1 sufficiently small and central. Up to a point that is true. But beyond that point there is a surprise. Further reducing the size of the holes does not reduce further the inaccuracy of the gun, but increases it. The pattern built up, by pulsing the gun many times and photographically recording the electron flashes, is something like

Fig. 2. Pattern built up by many pulses of electron gun of Fig. 1.

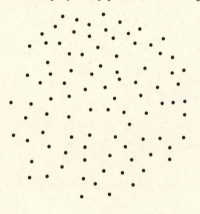

Fig. 3. Electron gun with two holes in second screen.

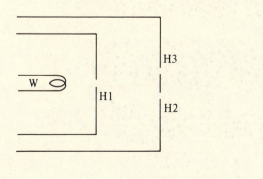

Fig. 2. The flashes are scattered over a region which gets bigger, rather than smaller, when the holes by which we try to determine the electron trajectory are reduced beyond a certain magnitude.

There is a still greater surprise when the hole H2 is replaced by two holes

Fig. 4. Guess, on basis of classical particle mechanics, for pattern built up by many pulses of electron gun of Fig. 3.

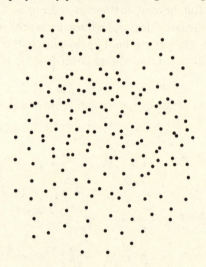

Fig. 5. Actual pattern from electron gun of Fig. 3.

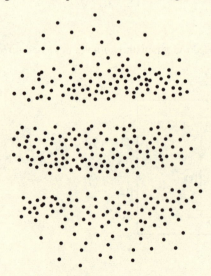

close together, Fig. 3. Instead of the contributions of these two holes just adding together, as in Fig. 4, an 'interference pattern' appears, as in Fig. 5. There are places on the screen that no electron can reach, when two holes are open, which electrons do reach when either hole alone is open. Although each electron passes through one hole or the other (or so we tend to think) it is as if the mere possibility of passing through the other hole influences its motion and prevents it going in certain directions. Here is the first hint of some queerness in the relation between possibility and actuality in quantum phenomena.

Forget for a moment that the patterns in Fig. 2 and Fig. 5 are built up from separated points (collected separately over a period of time) and look only at the general impression. Then these patterns become reminiscent of those which occur in classical physics in connection not with particles but with waves. Consider for example a regular train of waves on the surface of water. When they fall on a barrier with a hole, Fig. 6, they proceed more or less straight on, on the other side, when the hole is large compared with the wavelength. But when the hole is smaller, they diverge after passing through, Fig. 7, and to a degree which is greater the smaller the hole. This is

Fig. 6. Propagation of waves through hole much larger than wavelength.

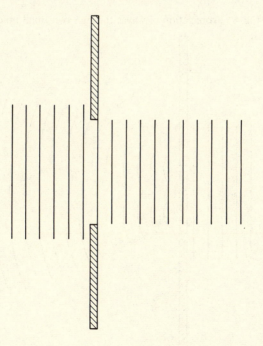

Fig. 7. Propagation of waves through hole much smaller than wavelength.

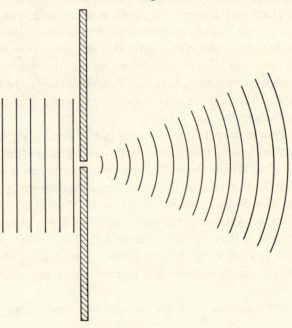

Fig. 8. Propagation of waves through two small holes.

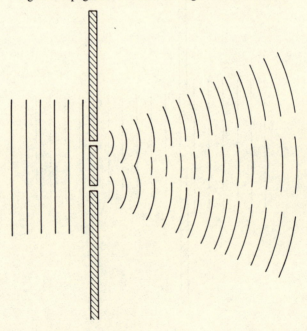

called 'wave diffraction'. And when the barrier has two small holes, Fig. 8, there are places behind the barrier where the surface of the water is undisturbed with both holes open, but disturbed when either separately is open. These are places where the waves from one hole try to raise the surface of the water while the waves from the other hole are trying to lower it, and vice versa. This is called 'wave interference'.

Returning to the electron then, we cannot tell in advance at just which point on the screen it will flash. But it seems that the places where it is likely to turn up are just those which a certain wave motion can appreciably reach.

It is the mathematics of this wave motion, which somehow controls the electron, that is developed in a precise way in quantum mechanics. Indeed the most simple and natural of the various equivalent ways in which quantum mechanics can be presented is called just 'wave mechanics'. What is it that 'waves' in wave mechanics? In the case of water waves it is the surface of the water that waves. With sound waves the pressure of the air oscillates. Light also was held to be a wave motion in classical physics. We were already a little vague about what was waving in that case... and even about whether the question made sense. In the case of the waves of wave mechanics we have no idea what is waving... and do not ask the question. What we do have is a mathematical recipe for the propagation of the waves, and the rule that the probability of an electron being seen at a particular place when looked for there (e.g. by introducing a scintillation screen) is related to the intensity there of the wave motion.

In my opinion the following point cannot be emphasised too strongly. When we work out a problem in wave mechanics, for example that of the precise performance of the electron gun, our mathematics is entirely concerned with waves. There is no hint in the mathematics of particles or particle trajectories. With the electron gun the calculated wave extends smoothly over an extended portion of the screen. There is no hint in the mathematics that the actual phenomenon is a minute flash at some particular point in that extended region. And it is only in applying the rule, relating the probable location of the flash to the intensity of the wave, that indeterminism enters the theory. The mathematics itself is smooth, deterministic, 'classical' mathematics... of classical waves.

So far it was only the single electron, proceeding from the hole H2 to the detection screen G, that was replaced by a wave in the mathematics. The screen G, in particular, was not discussed at all. It was simply assumed to have the capacity to scintillate. Suppose we wish to explain this capacity. Suppose we wish to calculate the intensity, the colour, or indeed the size of the scintillation (for it is not really a point)? We see that our treatment of the electron gun so far is neither complete nor accurate. If we wish to say more, and be more accurate, about its performance, then we have to see it as made of atoms, of electrons and nuclei. We have to apply to these entities the only mechanics that we know to be applicable... wave mechanics. Pursuing this line of thought, we are led, in the quest for more accuracy and completeness, to include more and more of the world in the wavy quantum mechanical 'system'... the photographic plate that records the scintillations, the developing chemicals that produce the photographic image, the eye of the observer...

But we cannot include the whole world in this wavy part. For the wave of the world is no more like the world we know than the extended wave of the single electron is like the tiny flash on the screen. We must always exclude part of the world from the wavy 'system', to be described in a 'classical' 'particulate' way, as involving definite events rather than just wavy possibilities. The purpose of the wave calculus is just that it yields formulae for probabilities of events at this 'classical' level.

Thus in contemporary quantum theory it seems that the world must be divided into a wavy 'quantum system', and a remainder which is in some sense 'classical'. The division is made one way or another, in a particular application, according to the degree of accuracy and completeness aimed at. For me it is the indispensibility, and above all the shiftiness, of such a division that is the big surprise of quantum mechanics. It introduces an essential ambiguity into fundamental physical theory, if only at a level of accuracy and completeness beyond any required in practice. It is the toleration of such an ambiguity, not merely provisionally but permanently, and at the most fundamental level, that is the real break with the classical ideal. It is this rather than the failure of any particular concept such as 'particle' or 'determinism'. In the remainder of this essay I will outline a number of world views which physicists have entertained in trying to digest this situation.

First, and foremost, is the purely pragmatic view. As we probe the world in regions remote from ordinary experience, for example the very big or the very small, we have no right to expect that familiar notions will work. We have no right to insist on concepts like space, time, causality, or even perhaps unambiguity. We have no right whatever to a clear picture of what goes on at the atomic level. We are very lucky that we can form rules of calculation, those of wave mechanics, which work. It is true that in principle there is some ambiguity in the application of these rules, in deciding just how the world is to be divided into 'quantum system' and the 'classical' remainder. But this matters not at all in practice. When in doubt, enlarge the quantum system. Then it is found that the division can be so made that moving it further makes very little difference to practical predictions. Indeed good taste and discretion, born of experience, allow us largely to forget, in most calculations, the instruments of observation. We can usually concentrate on a quite minute 'quantum system', and yet come up with predictions meaningful to experimenters who must use macroscopic instruments. This pragmatic philosophy is, I think, consciously or unconsciously the working philosophy of all who work with quantum theory in a practical way... when so working. We differ only in the degree of concern or complacency with which we view... out of working hours, so to speak... the intrinsic ambiguity in principle of the theory.

Niels Bohr, among the very greatest of theoretical physicists, made immense contributions to the development of practical quantum theory. And when this took definitive form, in the years following 1925, he was foremost in clarifying the way in which the theory should be applied to avoid contradictions at the practical level. No one more than he insisted that part of the world (indeed the vastly bigger part) must be held outside the 'quantum system' and described in classical terms. He emphasized that at this classical level we are concerned, as regards the present and the past, with definite events rather than wavy potentialities. And that at this level ordinary language and logic are appropriate. And that it is to statements in this ordinary language and logic that quantum mechanics must lead, however esoteric the recipe for generating these statements.

However Bohr went further than pragmatism, and put forward a philosophy of what lies behind the recipes. Rather than being disturbed by the ambiguity in principle, by the shiftiness of the division between 'quantum system' and 'classical apparatus', he seemed to take satisfaction in it. He seemed to revel in the contradictions, for example between 'wave' and 'particle', that seem to appear in any attempt to go beyond the pragmatic level. Not to resolve these contradictions and ambiguities, but rather to reconcile us to them, he put forward a philosophy which he called 'complementarity'. He thought that 'complementarity' was important not only for physics, but for the whole of human knowledge. The justly immense prestige of Bohr has led to the mention of complementarity in most text books of quantum theory. But usually only in a few lines. One is tempted to suspect that the authors do not understand the Bohr philosophy sufficiently to find it helpful. Einstein himself had great difficulty in reaching a sharp formulation of Bohr's meaning. What hope then for the rest of us? There is very little I can say about 'complementarity'. But I wish to say one thing. It seems to me that Bohr used this word with the reverse of its usual meaning. Consider for example the elephant. From the front she is head, trunk, and two legs. From the back she is bottom, tail, and two legs. From the sides she is otherwise, and from top and bottom different again. These various views are complementary in the usual sense of the word. They supplement one another, they are consistent with one another, and they are all entailed by the unifying concept 'elephant'. It is my impression that to suppose Bohr used the word 'complementary' in this ordinary way would have been regarded by him as missing his point and trivializing his thought. He seems to insist rather that we must use in our analysis elements which *contradict* one another, which do not add up to, or derive from, a whole. By 'complementarity' he meant, it seems to me, the reverse: contradictariness. Bohr seemed to like aphorisms such as: 'the opposite of a deep truth is also a deep truth': 'truth and clarity are complementary'. Perhaps he took a subtle satisfaction in the use of a familiar word with the reverse of its familiar meaning.

'Complementarity' is one of what might be called the 'romantic' world views inspired by quantum theory. It emphasizes the bizarre nature of the quantum world, the inadequacy of everyday notions and classical concepts. It lays stress on how far we have left behind naive 19th century materialism. I will describe two other romantic pictures, but will preface each by related unromantic notions.

Suppose that we accept Bohr's insistence that the very small and the very big must be described in very different ways, in quantum and classical terms respectively. But suppose we are sceptical about the possibility of such a division being sharp, and above all about the possibility of such a division being shifty. Surely the big and the small should merge smoothly with one another? And surely in fundamental physical theory this merging should be described not just by vague words but by precise mathematics? This mathematics would allow electrons to enjoy the cloudiness of waves, while allowing tables and chairs, and ourselves, and black marks on photographs, to be rather definitely in one place rather than another, and to be described in 'classical terms'. The necessary technical theoretical development involves introducing what is called 'nonlinearity', and perhaps what is called 'stochasticity', into the basic 'Schrödinger equation'. There have been interesting pioneer efforts in this direction, but not yet a breakthrough. This possible way ahead is unromantic in that it requires mathematical work by theoretical physicists, rather than interpretation by philosophers, and does not promise lessons in philosophy for philosophers.

There is a romantic alternative to the idea just mentioned. It accepts that the 'linear' wave mechanics does not apply to the whole world. It accepts that there is a division, whether sharp or smooth, between 'linear' and 'nonlinear', between 'quantum' and 'classical'. But instead of putting this division somewhere between small and big, it puts it between 'matter' (so to speak) and 'mind'. When we try to complete as far as possible the quantum theoretic account of the electron gun, we include first the scintillation screen, and then the photographic film, and then the developing chemicals, and then the eye of the experimenter... and then (why not) her brain. For the brain is made of atoms, of electrons and nuclei, and so why should we hesitate to apply wave mechanics... at least if we were smart enough to do the calculations for such a complicated assembly of atoms? But beyond the brain is... the mind. Surely the mind is not material? Surely here at last we come to something which is distinctly different from the glass screen, and the gelatine film... Surely it is here that we must expect some very different mathematics, (if mathematics at all), to be relevant? This view, that the necessary 'classical terms', and nonlinear mathematics, are in the mind, has been entertained especially by E. P. Wigner. And no one more eloquently than J. A. Wheeler has proposed that the very existence of the 'material' world may depend on the participation of mind. Unfortunately it has not yet been possible to develop these ideas in a precise way.

The last unromantic picture that I will present is the 'pilot wave' picture. It is due to de Broglie (1925) and Bohm (1952). While the founding fathers agonized over the question

'particle' or 'wave'

de Broglie in 1925 proposed the obvious answer

'particle' and 'wave'.

Is it not clear from the smallness of the scintillation on the screen that we have to do with a particle? And is it not clear, from the diffraction and interference patterns, that the motion of the particle is directed by a wave? De Broglie showed in detail how the motion of a particle, passing through just one of two holes in screen, could be influenced by waves propagating through both holes. And so influenced that the particle does not go where the waves cancel out, but is attracted to where they cooperate. This idea seems to me so natural and simple, to resolve the wave–particle dilemma in such a clear and ordinary way, that it is a great mystery to me that it was so generally ignored. Of the founding fathers, only Einstein thought that de Broglie was on the right lines. Discouraged, de Broglie abandoned his picture for many years. He took it up again only when it was rediscovered, and more systematically presented, in 1952, by David Bohm. In particular

Bohm developed the picture for many particles instead of just one. The generalization is straightforward. There is no need in this picture to divide the world into 'quantum' and 'classical' parts. For the necessary 'classical terms' are available already for individual particles (their actual positions) and so also for macroscopic assemblies of particles.

The de Broglie–Bohm synthesis, of particle and wave, could be regarded as a precise illustration of Bohr's complementarity... if Bohr had been using this word in the ordinary way. This picture combines quite naturally both the waviness of electron diffraction and interference patterns, and the smallness of individual scintillations, or more generally the definite nature of large scale happenings. The de B–B picture is also, by the way, quite deterministic. The initial configuration of the combined wave–particle system completely fixes the subsequent development. That we cannot predict just where a particular electron will scintillate on the screen is just because we cannot know everything. That we cannot arrange for impact at a chosen place is just because we cannot control everything.

We come finally to the romantic counterpart of the pilot wave picture. This is the 'many world interpretation', or MWI. It is surely the most bizarre of all the ideas that have come forth in this connection. It is most easily motivated, it seems to me, as a response to a central problem of the pragmatic approach... the so-called 'reduction of the wavefunction'. In discussing the electron gun, I emphasized the contrast between the extension of the wave and the minuteness of the individual flash. What happens to the wave where there is no flash? In the pragmatic approach the parts of the wave where there is no flash are just discarded... and this is effected by rule of thumb rather than by precise mathematics. In the pilot wave picture the wave, while influencing the particle, is not influenced by the particle. Flash or no flash, the wave just continues its mathematical evolution... even where it is 'empty' (very roughly speaking). In the MWI also the wave continues its mathematical way, but the notion of 'empty wave' is avoided. It is avoided by the assertion that everywhere that there *might* be a flash... there is a flash. But how can this be, for with one electron surely we see only one flash, at only one of the possible places? It can be because the world multiplies! After the flash there are as many worlds (at least) as places which can flash. In each world the flash occurs at just one place, but at different places in different worlds. The set of actual worlds taken together corresponds to all the possibilities latent in the wave. Quite generally, whenever there is doubt about what can happen, because of quantum uncertainty, the world multiplies so that all possibilities are actually realized. Persons of course multiply with the world, and those in any particular branch world experience only what happens in that branch. With one electron, each of us sees only one flash.

The MWI was invented by H. Everett in 1957. It has been advocated by such distinguished physicists as J. A. Wheeler, B. de Witt, and S. Hawking. It seems to attract especially quantum cosmologists, who wish to consider the world as a whole, and as a single quantum system, and so are particularly embarrassed by the requirement, in the pragmatic approach, for a 'classical' part outside the quantum system... i.e. outside the world. But this problem is already solved by the 'pilot wave' picture. It needs no extra classical part, for 'classical terms' are already applicable to the electron itself, and so to large assemblies of particles. The authors in question probably did not know this. For the pilot wave interpretation was rather deeply consigned to oblivion by the founding fathers, and by the writers of text-books.

The MWI is sometimes put forward as a working out of the hypothesis: the wavefunction is everything, there is nothing else. (Then the parts of the wavefunction cannot be distinguished from one another on the grounds of corresponding to possibility rather than actuality.) But here the authors, in my opinion, are mistaken. The MWI does add something to the wavefunction. I stressed in discussing the electron gun that the extended wave has little resemblance to the minute flash. Inspection of the wave itself gives no hint that the experienced reality is a scintillation... rather than, for example, an extended glow of unpredicted colour. That is to say, the extended wave does not simply fail to specify one of the possibilities as actual... it fails to list the possibilites. When the MWI postulates the existence of many worlds in each of which the photographic plate is blackened at particular position, it adds, surreptitiously, to the wavefunction, the missing classification of possibilities. And it does so in an imprecise way, for the notion of the position of a black spot (it is not a mathematical point), and indeed the concept of the reading of any macroscope instrument, is not mathematically sharp. One is given no idea of how far down towards the atomic scale the splitting of the world into branch worlds penetrates.

There then are six possible worlds to choose from, designed to accommodate the quantum phenomena. It would be possible to devize hybrids between them and maybe other worlds that are entirely different. I have tried to present them with some detachment, as if I did not regard one more than another to be pure fiction. I will now permit myself to express some personal opinions.

It is easy to understand the attraction of the three romantic worlds for journalists, trying to hold the attention of the man in the street. The opposite of a truth is also a truth! Scientists say that matter is not possible without mind! All possible worlds are actual worlds! Wow! And the journalists can write these things with good consciences, for things like this have indeed been said... out of working hours... by great physicists. For my part, I never got the hang of complementarity, and remain unhappy about contradictions. As regards mind, I am fully convinced that it has a central place in the ultimate nature of reality. But I am very doubtful that contemporary physics has reached so deeply down that that idea will soon be professionally fruitful. For our generation I think we can more profitably seek Bohr's necessary 'classical terms' in ordinary macroscopic objects, rather than in the mind of the observer. The 'many world interpretation' seems to me an extravagant, and above all an extravagantly vague, hypothesis. I could almost dismiss it as silly. And yet... It may have something distinctive to say in connection with the 'Einstein Podolsky Rosen puzzle', and it would be worthwhile, I think, to formulate some precise version of it to see if this is really so. And the existence of all possible worlds may make us more comfortable about the existence of our own world... which seems to be in some ways a highly improbable one.

The unromantic, 'professional', alternatives make much less good copy. The pragmatic attitude, because of its great success and immense continuing fruitfulness, must be held in high respect. Moreover it seems to me that in the course of time one may find that because of technical pragmatic progress the 'Problem of Interpretation of Quantum Mechanics' has been encircled. And the solution, invisible from the front, may be seen from the back. For the present, the problem is there, and some of us will not be able to resist paying attention to it. The nonlinear Schrödinger equation seems to me to be the best hope for a precisely formulated theory which is very close to the pragmatic version. But while we get along so well without precision, the pragmatists are not going to help to develop it. The 'pilot wave' picture is an almost trivial reconciliation of quantum phenomena with the classical ideals of theoretical physics... a closed set of equations, whose solutions are to be taken seriously, and not mutilated ('reduced') when embarrassing. However it would be wrong to leave the reader with the impression that, with the pilot wave picture, quantum theory simply emerges into the light of day, with the transparency of pure water. The very clarity of this picture puts in evidence the extraordinary 'non-locality' of quantum theory. But that is another story.

To what extent are these possible worlds fictions? They are like literary fiction in that they are free inventions of the human mind. In theoretical physics sometimes the inventor knows from the beginning that the work is fiction, for example when it deals with a simplified world in which space has only one or two dimensions instead of three. More often it is not known till later, when the hypothesis has proved wrong, that fiction is involved. When being serious, when not exploring deliberately simplified models, the theoretical physicist differs from the novelist in thinking that maybe the story might be true. Perhaps there is some analogy with the historical novelist. If the action is put in the year 1327, the Pope must be located in Avignon, not Rome. The serious theories of theoretical physicists must not contradict experimental facts. If thoughts are put into the mind of Pope John XXII, then they must be reasonably consistent with what is known of his words and actions. When we invent worlds in physics we would have them to be mathematically consistent continuations of the visible world into the invisible... even when it is beyond human capability to decide which, if any, of those worlds is the true one. Literary fiction, historical or otherwise, can be professionally good or bad (I think). We could also consider how our possible worlds in physics measure up to professional standards. In my opinion the pilot wave picture undoubtedly shows the best craftsmanship among the pictures we have considered. But is that a virtue in our time?

21

EPR correlations and EPW distributions

Dedicated to Professor E. P. Wigner

It is known that with Bohm's example of EPR correlations, involving particles with spin, there is an irreducible non-locality. The non-locality cannot be removed by the introduction of hypothetical variables unknown to ordinary quantum mechanics. How is it with the original EPR example involving two particles of zero spin? Here we will see that the Wigner phase space distribution[1] illuminates the problem.

Of course, if one admits 'measurement' of arbitrary 'observables' on arbitrary states, it is easy to mimic[2] the EPRB situation. Some steps have been made towards realism in that connection[3]. Here we will consider a narrower problem, restricted to 'measurement' of positions only, on two non-interacting spinless particles in free space. EPR considered 'measurement' of momenta as well as positions. But the simplest way to 'measure' the momenta of free particles is just to wait a long time and 'measure' their positions. Here we will allow position measurements at arbitrary times t_1 and t_2 on the two particles respectively. This corresponds to 'measuring' the combinations

$$\hat{q}_1 + t_1 \hat{p}_1/m_1, \quad \hat{q}_2 + t_2 \hat{p}_2/m_2 \tag{1}$$

at time zero, where m_1 and m_2 are the masses, and the $\hat{q}$ and $\hat{p}$ are position and momentum operators. We will be content here with just one space dimension.

The times t_1 and t_2 play the same roles here as do the two polarizer settings in the EPRB example. One can envisage then some analogue of the CHHS inequality[4,5] discriminating between quantum mechanics on the one hand and local causality on the other.

The QM probability of finding, at times t_1 and t_2 respectively, the particles at positions q_1 and q_2 respectively, is

$$\rho(q_1, q_2, t_1, t_2)$$

with

$$\rho = |\psi(q_1, q_2, t_1, t_2)|^2 \tag{2}$$

The two-time wave function ψ satisfies the two Schrödinger equations

$$\left.\begin{aligned} i\hbar\partial\psi/\partial t_1 &= H_1\psi = (\hat{p}_1^2/2m_1)\psi \\ i\hbar\partial\psi/\partial t_2 &= H_2\psi = (\hat{p}_2^2/2m_2)\psi \end{aligned}\right\} \tag{3}$$

with

$$i\hat{p}_1 = \hbar\partial/\partial q_1, \quad i\hat{p}_2 = \hbar\partial/\partial q_2$$

For simplicity we will consider the case of equal masses, and take units such that

$$m_1 = m_2 = \hbar = 1$$

The same ρ, (2), can be obtained from the corresponding two-time Wigner distribution:

$$\rho = \iint \frac{\mathrm{d}p_1}{2\pi}\frac{\mathrm{d}p_2}{2\pi}\, W(q_1,q_2,p_1,p_2,t_1,t_2) \tag{4}$$

where

$$\left.\begin{aligned} W = \iint \mathrm{d}y_1\,\mathrm{d}y_2\, e^{-i(p_1 y_1 + p_2 y_2)} \psi\left(q_1 + \frac{y_1}{2}, q_2 + \frac{y_2}{2}, t_1, t_2\right) \\ \cdot\psi^*\left(q_1 - \frac{y_1}{2}, q_2 - \frac{y_2}{2}, t_1, t_2\right) \end{aligned}\right\} \tag{5}$$

From (3),

$$(\partial/\partial t_1 + p_1\partial/\partial q_1)W = (\partial/\partial t_2 + p_2\partial/\partial q_2)W = 0 \tag{6}$$

That is, W evolves exactly as does a probability distribution for a pair of freely-moving classical particles:

$$W(q_1,q_2,p_1,p_2,t_1,t_2) = W(q_1 - p_1 t_1, q_2 - p_2 t_2, p_1, p_2, t_1, t_2) \tag{7}$$

When W happens to be initially nowhere negative, the classical evolution (7) preserves the non-negativity. The original EPR wave function[6]

$$\delta((q_1 + \tfrac{1}{2}q_0) - (q_2 - \tfrac{1}{2}q_0)), \tag{8}$$

assumed to hold at $t_1 = t_2 = 0$, gives

$$W(q_1,q_2,p_1,p_2,0,0) = \delta(q_1 - q_2 + q_0)2\pi\delta(p_1 + p_2) \tag{9}$$

This is nowhere negative, and the evolved function (7) has the same property. Thus in this case the EPR correlations are precisely those between two classical particles in independent free classical motion.

With the wave function (8), then, there is no non-locality problem when the incompleteness of the wave function description is admitted. The Wigner distribution provides a local classical model of the correlations. Since the Wigner distribution appeared in 1932, this remark could already have been made in 1935. Perhaps it was. And perhaps it was already anticipated that wave functions, other than (8), with Wigner distributions that are not non-negative, would provide a more formidable problem. We will see that this is so.

Consider, for example, the initial wave function

$$(q^2 - 2a^2)e^{-q^2/(2a^2)} \tag{10}$$

where

$$q = (q_1 + q_0/2) - (q_2 - q_0/2) \tag{11}$$

It could be made normalizable by including a factor

$$\exp - ((q_1 + q_0/2) + (q_2 - q_0/2))^2/(2b^2) \tag{12}$$

But we will immediately anticipate the limit $b \to \infty$, and will consider only relative probabilities. Choosing the unit of length so that $a = 1$ gives as the initial Wigner distribution

$$W(q_1, q_2, p_1, p_2, 0, 0) = Ke^{-q^2}e^{-p^2}\{(q^2 + p^2)^2 - 5q^2 + p^2 + 11/4\}\delta(p_1 + p_2) \tag{13}$$

where K is an unimportant constant, and

$$p = (p_1 - p_2)/2 \tag{14}$$

This W, (13), is in some regions negative, for example at $(p = 0, q = 1)$. It no longer provides an explicitly local classical model of the correlations. I do not know that the failure of W to be non-negative is a *sufficient* condition in general for a locality paradox. But it happens that (13) implies, as well as negative regions in the Wigner distribution, a violation of the CHHS locality inequality.

To see this, first calculate the two-time position probability distribution, either from (4), (7) and (13), or from (2) and the solution of (3). The result is

$$\rho = K'(1 + \tau^2)^{-5/2}\{q^4 + q^2(2\tau^2 - 4) + 3(1 + \tau^2) + (1 + \tau^2)^2\}e^{-q^2/1 + \tau^2} \tag{15}$$

where K' is an unimportant constant, and

$$\tau = t_1 + t_2 \tag{16}$$

Calculate then the probability D that $(q_1 + q_0/2)$ and $(q_2 - q_0/2)$ disagree in sign:

$$D(t_1, t_2) = \int_{\tau\infty}^{\infty} dq |q| \rho \tag{17}$$

$$= K''(\tau^2 + \tfrac{2}{5})/\sqrt{\tau^2 + 1} \tag{18}$$

Consider finally the CHHS inequality

$$E(t_1, t_2) + E(t_1, t'_2) + E(t'_1, t_2) - E(t'_1, t'_2) \leqslant 2 \tag{19}$$

where

$$\left. \begin{array}{l} E(t_1, t_2) = \text{probability of } (+, +) + \text{probability of } (-, -) \\ \qquad - \text{probability of } (+, -) - \text{probability of } (-, +) \end{array} \right\} \tag{20}$$

$$= 1 - 2 \left(\text{probability}(+, -) + \text{probability}(-, +) \right) \tag{21}$$

Using (21), (19) becomes

$$D(t_1, t_2) + D(t_1, t'_2) + D(t'_1, t_2) - D(t'_1, t'_2) \geqslant 0 \tag{22}$$

With

$$t'_1 = 0, \quad t_2 = \tau, \quad t_1 = -2\tau, \quad t'_2 = 3\tau \tag{23}$$

and assuming (in view of (18))

$$D(t_1, t_2) = F(|t_1 + t_2|) \tag{24}$$

(22) gives (for τ positive)

$$3F(\tau) - F(3\tau) \geqslant 0 \tag{25}$$

But this is violated by (18) when τ exceeds about 1. There is a real non-locality problem with the wave function (10).

Only some epsilonics will be added here. The essential assumption leading to (19) is (roughly speaking) that measurement on particle 1 is irrelevant for particle 2, and vice versa. This follows from local causality[7] if we look for the particles only in limited space-time regions

$$\begin{array}{ll} |q_1 + q_0/2| < L, & |t_1| < T \\ |q_2 - q_0/2| < L, & |t_2| < T \end{array} \tag{26}$$

with

$$L \ll q_0, \quad cT \ll q_0 \tag{27}$$

so that the two regions (26) have spacelike separation. We must, however, make L large enough, compared with b in (12), so that the particles are almost sure to be found in the regions in question, for in passing from (20) to (21) it was assumed that the four probabilities in (20) add to unity; and b in turn must be large compared with a, as was used to simplify the detailed calculations. So as well as (27), we specify.

$$1 \gg a/b \gg (b/L)e^{-L^2/b^2} \qquad (28)$$

Notes and references

1 E. P. Wigner, *Phys. Rev.* **40**, 749 (1932).

2 J. S. Bell, *Physics* **1**, 195 (1965).

3 M. A. Horne and A. Zeilinger, in *Symposium on the Foundations of Modern Physics, Joensuu 1985*. Eds. P. Lahti and P. Mittelstaedt, World Scientific, Singapore (1985). And 9 below.

4 J. F. Clauser, R. A. Holt, M. A. Horne and A. Shimony, *Phys. Rev. Lett.* **23**, 880 (1969).

5 J.F. Clauser and A. Shimony, *Rep. Prog. Phys.* **41**, 1881 (1978).

6 A. Einstein, B. Podolsky and N. Rosen, *Phys. Rev.* **47**, 779 (1935).

7 J. S. Bell, *Theory of Local Beables*, preprint CERN-TH 2053/75, reprinted in *Epistemological Letters* **9**, 11 (1976), and in *Dialectica* **39**, 86 (1985). The notion of local causality presented in this reference involves complete specification of the beables in an infinite space-time region. The following conception is more attractive in this respect: In a locally-causal theory, probabilities attached to values of local beables in one space-time region, when values are specified for *all* local beables in a second space-time region fully obstructing the backward light cone of the first, are unaltered by specification of values of local beables in a third region with spacelike separation from the first two.

8 The discussion has a new interest when the positions q_1 and q_2 are granted beable status. Then we can consider their actual values rather than 'measurement results', at arbitrary times t_1 and t_2. External intervention by hypothetically free-willed experimenters is not involved.

9 See also L. A. Khalfin and B. S. Tsirelson, in the Joensuu proceedings (Ref. 3), and A. M. Cetto, L. de la Peña and E. Santos, *Phys. Lett.* **A113**, 304 (1985). These last authors invoke the Wigner distribution.

22

Are there quantum jumps?

If we have to go on with these damned quantum jumps, then I'm sorry that I ever got involved. E. Schrödinger

1 Introduction

I have borrowed the title of a characteristic paper by Schrödinger (Schrödinger, 1952). In it he contrasts the smooth evolution of the Schrödinger wavefunction with the erratic behaviour of the picture by which the wavefunction is usually supplemented, or 'interpreted', in the minds of most physicists. He objects in particular to the notion of 'stationary states', and above all to 'quantum jumping' between those states. He regards these concepts as hangovers from the old Bohr quantum theory, of 1913, and entirely unmotivated by anything in the mathematics of the new theory of 1926. He would like to regard the wavefunction itself as the complete picture, and completely determined by the Schrödinger equation, and so evolving smoothly without 'quantum jumps'. Nor would he have 'particles' in the picture. At an early stage, he had tried to replace 'particles' by wavepackets (Schrödinger, 1926). But wavepackets diffuse. And the paper of 1952 ends, rather lamely, with the admission that Schrödinger does not see how, for the present, to account for particle tracks in track chambers... nor, more generally, for the definiteness, the particularity, of the world of experience, as compared with the indefiniteness, the waviness, of the wavefunction. It is the problem that he had had (Schrödinger, 1935a) with his cat. He thought that she could not be both dead and alive. But the wavefunction showed no such commitment, superposing the possibilities. Either the wavefunction, as given by the Schrödinger equation, is not everything, or it is not right.

Of these two possibilities, that the wavefunction is not everything, or not right, the first is developed especially in the de Broglie–Bohm 'pilot wave' picture. Absurdly, such theories are known as 'hidden variable' theories. Absurdly, for there it is not in the wavefunction that one finds an image of the visible world, and the results of experiments, but in the complementary 'hidden'(!) variables. Of course the extra variables are not confined to the visible 'macroscopic' scale. For no sharp definition of such a scale could be made. The 'microscopic' aspect of the complementary variables is indeed hidden from us. But to admit things not visible to the gross creatures that we are is, in my opinion, to show a decent humility, and not just a lamentable addiction to metaphysics. In any case, the most hidden of all variables, in the pilot wave picture, is the wavefunction, which manifests itself to us only by its influence on the complementary variables.

If, with Schrödinger, we reject extra variables, then we must allow that his equation is not always right. I do not know that he contemplated this conclusion, but it seems to me inescapable. Anyway it is the line that I will follow here. The idea of a small change in the mathematics of the wavefunction, one that would little affect small systems, but would become important in large systems, like cats and other scientific instruments, has often been entertained. It seems to me that a recent idea (Ghirardi, Rimini and Weber, 1985), a specific form of spontaneous wavefunction collapse, is particularly simple and effective. I will present it below. Then I will consider what light it throws on another of Schrödinger's preoccupations. He was one of those who reacted most vigorously (Schrödinger, 1935*a*, *b*, 1936) to the famous paper of Einstein, Podolsky and Rosen (1935). As regards what he called 'quantum entanglement', and the resulting EPR correlations, he 'would not call that *one* but rather *the* characteristic trait of quantum mechanics, the one that enforces its entire departure from classical lines of thought'.

2 Ghirardi, Rimini and Weber

The proposal of Ghirardi, Rimini and Weber, is formulated for non-relativistic Schrödinger quantum mechanics. The idea is that while a wavefunction

$$\psi(t, \mathbf{r}_1, \mathbf{r}_2, \ldots, \mathbf{r}_N) \tag{1}$$

normally evolves according to the Schrödinger equation, from time to time it makes a jump. Yes, a jump! But we will see that these GRW jumps have little to do with those which Schrödinger objected so strongly. The only resemblance is that they are random and spontaneous. The probability per unit time for a GRW jump is

$$\frac{N}{\tau}, \tag{2}$$

where N is the number of arguments $\mathbf{r}$ in the wavefunction, and τ is a new constant of nature. The jump is to a 'reduced' or 'collapsed' wavefunction

$$\psi' = \frac{j(\mathbf{x} - \mathbf{r}_n)\psi(t, \ldots)}{R_n(\mathbf{x})}, \tag{3}$$

where $\mathbf{r}_n$ is randomly chosen from the arguments $\mathbf{r}$. The jump factor j is normalized:

$$\int d^3 \mathbf{x} |j(\mathbf{x})|^2 = 1. \tag{4}$$

Ghirardi, Rimini and Weber suggest a Gaussian:

$$j(\mathbf{x}) = K \exp(-\mathbf{x}^2 / 2a^2) \tag{5}$$

where a is again a new constant of nature. R is a renormalization factor:

$$|R_n(\mathbf{x})|^2 = \int d^3 \mathbf{r}_1 \cdots d^3 \mathbf{r}_N |j\psi|^2. \tag{6}$$

Finally the collapse centre $\mathbf{x}$ is randomly chosen with probability distribution

$$d^3 \mathbf{x} |R_n(\mathbf{x})|^2. \tag{7}$$

For the new constants of nature, GRW suggest as orders of magnitude

$$\tau \approx 10^{15} \, \mathrm{s} \approx 10^8 \, \mathrm{year} \tag{8}$$

$$a \approx 10^{-5} \, \mathrm{cm}. \tag{9}$$

An immediate objection to the GRW spontaneous wavefunction collapse is that it does not respect the symmetry or antisymmetry required for 'identical particles'. But this will be taken care of when the idea is developed in the field theory context, with the GRW reduction applied to 'field variables' rather than 'particle positions'. I do not see why that should not be possible, although novel renormalization problems may arise.

There is no problem in dealing with 'spin'. The wavefunctions ψ and ψ' in (3) can be supposed to carry suppressed spin indices.

Consider now the wavefunction

$$\phi(\mathbf{s}_1 \cdots \mathbf{s}_L)\chi(\mathbf{r}_1 \cdots \mathbf{r}_M), \tag{10}$$

where L is not very big and M is very very big. The first factor, ϕ, might represent a small system, for example an atom or molecule, that is temporarily isolated from the rest of the world ... the latter, or part of it, represented by the second factor, χ. The GRW process for the complete wavefunction implies independent GRW processes for the two factors. From (8) we can forget about GRW processes in the small system. But in the big system, with M of order say 10^{20} or larger, the mean lifetime before a

GRW jump is some

$$\frac{10^{15}}{10^{20}} = 10^{-5}\,\text{s} \tag{11}$$

or less.

Consider next a wavefunction like

$$\phi_1(\mathbf{s}_1 \cdots \mathbf{s}_L)\chi_1(\mathbf{r}_1 \cdots \mathbf{r}_M) + \phi_2(\mathbf{s}_1 \cdots \mathbf{s}_L)\chi_2(\mathbf{r}_1 \cdots \mathbf{r}_M). \tag{12}$$

This might represent the aftermath of a 'quantum measurement' situation. Some 'property' of the small system has been 'measured' by interaction with a large 'instrument', which is thrown as a result into one or other of the states χ_1 or χ_2, corresponding to different pointer readings. This macroscopic difference between χ_1 and χ_2 implies that, for very many arguments $\mathbf{r}$, multiplication of the wavefunction by $j(\mathbf{x} - \mathbf{r})$ will reduce to zero one or other of the terms in (12). Thus in a time of order (11) one of the terms will disappear, and only the other will propagate. The wavefunction commits itself very quickly to one pointer reading or the other. Moreover, the probability that one term rather than the other survives is proportional to the fraction of the total norm which it carries – in agreement with the rule of pragmatic quantum theory.

Quite generally any embarrassing macroscopic ambiguity in the usual theory is only momentary in the GRW theory. The cat is not both dead and alive for more than a split second. One could worry perhaps if the GRW process does not go too far. In the usual pragmatic theory the 'reduction' or 'collapse' of the wavefunction is an operation performed by the theorist at some time convenient for her. Usually she will delay this till the Schrödinger equation has established a very big difference between χ_1 and χ_2. The GRW process is one of nature, and comes about as soon as the difference between χ_1 and χ_2 is big enough. I think that with suitable values of the natural constants (8, 9) the GRW theory will nevertheless agree with the pragmatic theory in practice. But studies on models would be useful to build up confidence in this.

3 Quantum entanglement

There is nothing in this theory but the wavefunction. It is in the wavefunction that we must find an image of the physical world, and in particular of the arrangement of things in ordinary three-dimensional space. But the wavefunction as a whole lives in a much bigger space, of $3N$-dimensions. It makes no sense to ask for the amplitude or phase or whatever of the wavefunction at a point in ordinary space. It has neither amplitude nor phase nor anything else until a multitude of points in ordinary three-space are specified. However, the GRW jumps (which are part of the wavefunction, not something else) are well localized in ordinary space. Indeed each is centred on a particular specetime point $(\mathbf{x}, t)$. So we can propose these events as the basis of the 'local beables' of the theory. These are the mathematical counterparts in the theory to real events at definite places and times in the real world (as distinct from the many purely mathematical constructions that occur in the working out of physical theories, as distinct from things which may be real but not localized, and as distinct from the 'observables' of other formulations of quantum mechanics, for which we have no use here). A piece of matter then is a galaxy of such events. As a schematic psychophysical parallelism we can suppose that our personal experience is more or less directly of events in particular pieces of matter, our brains, which events are in turn correlated with events in our bodies as a whole, and they in turn with events in the outer world.

In this paper we will use the notion of localization of events only in a rough way. We will localize them in one or other of two widely separated regions of space which we suppose to be occupied by two widely separated systems.

Let the arguments $\mathbf{s}$ and $\mathbf{r}$ in (12) refer to the two sides, respectively, in an Einstein–Podolsky–Rosen–Bohm setup, with L as well as M now large. A source, which for simplicity we omit from the analysis, emits a pair of spin $-\frac{1}{2}$ neutrons in the singlet spin state. They move through Stern–Gerlach magnets to counters which register for each neutron whether it has been deflected 'up' or 'down' in the corresponding magnet. According to the Schrödinger equation the wavefunction would come out like (12), with ϕ_1 or ϕ_2 corresponding to 'up' or 'down' on the left, and χ_1 or χ_2 corresponding to 'down' or 'up' on the right. Suppose that the left hand counters are closer to the source, and so register before the right hand ones. That is to say, suppose that ϕ_1 differs macroscopically from ϕ_2 before χ_1 from χ_2. Then the GRW jumps on the left quickly reduce the wavefunction to one or other of the two terms in (12). The choice between χ_1 and χ_2, as well as between ϕ_1 and ϕ_2, has then been made. The jumps on the left are decisive, and those on the right have no opportunity to be so.

In all this the GRW account is very close to that of a common way of presenting conventional quantum mechanics, with 'measurement' causing 'wavefunction collapse' – and with a 'measurement' somewhere causing 'collapse' everywhere. But it is important that in the GRW theory everything, including 'measurement', goes according to the mathematical equations of the theory. Those equations are not disregarded from time to time on the basic of supplementary, imprecise, verbal, prescriptions.

In this EPRB situation, an 'up' on the left implies a subsequent 'down' on the right, and vice versa. Now of course it was not the existence of correlations between distant events that scandalized EPR, and led Einstein (Einstein, 1949) to use the word 'paradox' in this connection. Such correlations are common in daily life. If I find that I have brought only one glove, the left handed, then I confidently predict that the one at home will be found to be right handed. In the everyday conception of things there is no puzzle here. Both gloves have been there all morning, and each has been right or left handed all the time. Observation of the one taken from my pocket gives information about, but does not influence, the one left at home. As regards EPRB correlations, what is disturbing about quantum mechanics, especially as sharpened by GRW, is that before the first 'measurement' there *is* nothing but the quantum mechanical wavefunction – entirely neutral between the two possibilities. The decision between these possibilities is made for both of the mutually distant systems only by the first 'measurement' on one of them. There is no question, if there *was* nothing but the wavefunction, of just revealing a decision already taken. It was this 'spooky action at a distance', the immediate determining of events in a distant system by events in a near system, that scandalized EPR. They concluded that quantum mechanics must, at best, be incomplete. There must be in nature additional variables, not yet known to quantum mechanics, in both systems, which determine in advance the results of experiments, and which happen to have become correlated at the source – just as gloves happen to be sold in matching pairs.

It is now very difficult to maintain this hope, that local causality might be restored to quantum mechanics by the addition of complementary variables. The perfect correlations actually considered by EPR, with parallel polarizers in the EPRB setup, do not present any difficulty in this respect. But the imperfect correlations implied by quantum mechanics, for misaligned polarizers, prove more intractable (e.g. Bell, 1981).

The GRW theory does not add variables. But by adding mathematical precision to the jumps in the wavefuction, it seems simply to make precise the action at a distance of ordinary quantum mechanics. The most disturbing aspect of this is the apparent difficulty of reconciling it with Lorentz invariance. For in a Lorentz invariant theory we tend to think that 'nothing goes faster than light'. So we turn now to a discussion of Lorentz invariance.

4 Relative time translation invariance

Of course we cannot discuss full Lorentz invariance in the context of the nonrelativistic model presented above. But there is a residue, or at least an analogue, of Lorentz invariance, which can be discussed in the case of two widely separated systems. Consider the Lorentz transformation

$$z' = \gamma(z - vt), \quad t' = \gamma(t - vz) \tag{13}$$

with x and y unchanged, where the velocity of light has been set equal to unity, and

$$\gamma = \frac{1}{(1 - v^2)^{1/2}}. \tag{14}$$

In the case of a system at a large distance, a, from the origin, it is convenient to introduce a new origin, so that

$$z \to z + a. \tag{15}$$

Then (13) becomes

$$z' = -a + \gamma(z + a - vt), \quad t' = \gamma(t - v(z + a)). \tag{16}$$

Taking v very small and a very large so that

$$va = k \tag{17}$$

(16) becomes

$$z' = z, \quad t' = t - k. \tag{18}$$

In the case of a single system this tells us simply to expect invariance with respect to translation in time. But in the case of two systems displaced from the origin in opposite directions, and so with different signs for k, it tells us to expect invariance with respect to displacement in *relative* time.

Multiple time formalism, with independent times for different particles, or for different points in space, is an old story in relativistic quantum theory. It is less familiar in the context of the nonrelativistic theory. However, it is easily implemented *in the case of noninteracting systems* at the level of the Schrödinger equation. Let two noninteracting subsystems have separate Hamiltonians A and B, respectively, so that the total Hamiltonian is

$$H = A + B. \tag{19}$$

Then from the ordinary one-time wavefunction $\psi(t, \ldots)$ we can define a two-time wavefunction

$$\psi(t', t'', \ldots) = \frac{\exp \mathrm{i}(t - t')A}{\hbar} \frac{\exp \mathrm{i}(t - t'')B}{\hbar} \psi(t, \ldots). \tag{20}$$

Since A and B commute, the relative order of the two exponentials in (20) is unimportant. (However, if A and B are time-dependent, the two exponentials must separately be time ordered, as in (A.5)). The two-time wavefunction satisfies the two Schrödinger equations

$$\frac{\hbar i \partial}{\partial t'} \psi(t', t'' \ldots) = A\psi(t', t'', \ldots) \tag{21}$$

$$\frac{\hbar i \partial}{\partial t''} \psi(t', t'' \ldots) = B\psi(t', t'', \ldots). \tag{22}$$

These equations are invariant against independent shifts in the origins of
the two time variables (provided any time dependent external fields in A
and B are shifted appropriately).

It remains to see if this relative time invariance survives the introduction
of the GRW jumps. It does. I did not find a short elegant argument, and
have relegated the clumsy arguments that I did find to an appendix. From
the ordinary one-time wavefunction for time i, a two-time wavefunction can
again be constructed. It incorporates the jumps of subsystem-1 between
times i and i', and those of subsystem-2 between i and i''. In terms of this a
formula can be found (A.22, A.23) for the probability of subsequent jumps
before times f' and f'' in the two subsystems respectively. It can be
interpreted as supplementing (21, 22) by giving the probabilities for jumps
in the two systems as t' and t'' are advanced independently from
independent starting points. It does not depend on t' or t'' except through
the two-time wavefunction ψ (and any time dependent external fields in
Hamiltonians A and B). The relative time translation invariance of the
theory is then manifest.

The reformulation (A.22, A.23) of the theory can also be used to calculate
the statistics of jumps in one system separately, disregarding what happens
in the other. The result (A.24, A.25) makes no reference to the second
system. Events in one system, considered separately, allow no inference
about events in the other, nor about external fields at work in the other, . . .
nor even about the very existence of the other system. There are no
'messages' in one system from the other. The inexplicable correlations of
quantum mechanics do not give rise to signalling between noninteracting
systems. Of course, however, there may be correlations (e.g. those of EPRB)
and if something about the second system is given (e.g. that it is the other
side of an EPRB setup) and something about the overall state (e.g. that it is
the EPRB singlet state) then inferences from events in one system (e.g. 'yes'
from the 'up' counter) to events in the other (e.g. 'yes' from the 'down'
counter) are possible.

5 Conclusion

I think that Schrödinger could hardly have found very compelling the
GRW theory as expounded here – with the arbitrariness of the jump

function, and the elusiveness of the new physical constants. But he might have seen in it a hint of something good to come. He would have liked, I think, that the theory is completely determined by the equations, which do not have to be talked away from time to time. He would have liked the complete absence of particles from the theory, and yet the emergence of 'particle tracks', and more generally of the 'particularity' of the world, on the macroscopic level. He might not have liked the GRW jumps, but he would have disliked them less than the old quantum jumps of his time. And he would not have been at all disturbed by their indeterminism. For as early as 1922, following his teacher Exner, he was expecting the fundamental laws to be statistical in character: '... once we have discarded our rooted predilection for absolute Causality, we shall succeed in overcoming the difficulties...' (Schrödinger, 1957).

For myself, I see the GRW model as a very nice illustration of how quantum mechanics, to become rational, requires only a change which is very small (on some measures!). And I am particularly struck by the fact that the model is as Lorentz invariant as it could be in the nonrelativistic version. It takes away the ground of my fear that any exact formulation of quantum mechanics must conflict with fundamental Lorentz invariance.

Appendix

Let

$$P(f; \mathbf{x}_m, n_m, t_m; \dots \mathbf{x}_1, n_1, t_1; i) \mathrm{d}^3 \mathbf{x}_1 \dots \mathrm{d}^3 \mathbf{x}_m \, \mathrm{d}t_1 \dots \mathrm{d}t_m \qquad (\mathrm{A}.1)$$

be the probability that between some time i and some later time f there are m jumps, with the first at time t_1 in the interval $\mathrm{d}t_1$, involving argument $\mathbf{r}_{n_1}$, and centred at $\mathbf{x}_1$ in $\mathrm{d}^3 \mathbf{x}_1$; and with the second at time t_2, involving argument $\mathbf{r}_{n_2}$, centred at $\mathbf{x}_2, \dots$ and so on. Then, from the basic assumptions,

$$P = \exp \lambda N(i - f) \langle i | E^+(f, i) E(f, i) | i \rangle, \qquad (\mathrm{A}.2)$$

where N is the total 'particle number', $|i\rangle$ denotes the initial state

$$|i\rangle = \psi(i, \mathbf{r}_1, \mathbf{r}_2 \dots) \qquad (\mathrm{A}.3)$$

and

$$E(f, i) = U(f, t_m) j(n_m, \mathbf{x}_m) \dots U(t_2, t_1) j(n_1, \mathbf{x}_1) U(t_1, i) \qquad (\mathrm{A}.4)$$

with

$$U(s, t) = T \exp \int_s^t \mathrm{d}t' \, \frac{H(t')}{i\hbar} \qquad (\mathrm{A}.5)$$

and

$$j(n, x) = \lambda^{1/2} j(\mathbf{x} - \mathbf{r}_n). \qquad (\mathrm{A}.6)$$

In (A.5) we allow that the Hamiltonian might be time dependent, and so have a time-ordered product. Note the unitarity relation

$$U^+U = 1. \tag{A.7}$$

The leftmost U in (A.4) is actually redundant in (A.2), because of (A.7), but it is convenient later. The exponential in front of (A.2) arises from a product of exponentials

$$\exp - \lambda N(t' - t),$$

which are the probabilities of having no jumps in the corresponding time intervals. The formulae could be simplified somewhat by introducing Heisenberg operators, but we will not do so here.

Let us calculate from (A.1)–(A.4), for given i, the conditional probability distribution for jumps in the interval i' till f when the jumps between i and i' are given. We have only to divide (A.1) by the probability for the given jumps:

$$\exp \lambda N(i - i')|R|^2 d^3 x_1 \ldots dt_1 \ldots \tag{A.8}$$

with, from (A.2),

$$|R|^2 = \langle i|E^+(i',i)E(i',i)|i\rangle. \tag{A.9}$$

The result may be expressed in terms of

$$|i'\rangle = \frac{E(i',i)|i\rangle}{R} \tag{A.10}$$

when we note the factorization property

$$E(f,i) = E(f,i')E(i',i). \tag{A.11}$$

If we renumber the jumps in the reduced interval after i' to begin again with 1, we find again just (A.1)–(A.4) with i replaced everywhere by i'. So this was only a rather elaborate consistency check. But the manipulations involved will be useful for another purpose in a moment.

Let us now calculate from (A.1)–(A.4), with fixed f, the probability P' for jumps specified only up to some earlier time f', regardless of what happens later. To do so we have to sum over all possibilities in the interval between f' and f. There might be 0, 1, 2, ... extra jumps in that remaining interval. The probability of the given jumps in the reduced interval, and no jumps in the remainder, is given directly by (A.2), which we rewrite as

$$X_0 \exp \lambda N(i - f')\langle i|E^+(f',i)E(f',i)|i\rangle \tag{A.12}$$

with
$$X_0 = \exp \lambda N(f' - f). \tag{A.13}$$

With one extra jump, E^+E in the expectation value is replaced by

$$E^+ U^+ |j(n,x)|^2 UE, \tag{A.14}$$

where the extra factor U evolves the system from time f' till the time t of the extra jump (n,x). Integration over x, using (4), replaces $|j(n,x)|^2$ by λ. The extra U^+U then goes away by unitarity. Summation over n gives a factor N, and integration over time t gives a factor $(f - f')$. Then the total one extra jump contribution to P' is (A.12) with X_0 replaced by

$$X_1 = \lambda N(f - f') \exp \lambda N(f' - f). \tag{A.15}$$

Proceeding in this way we find for the n-extra-jump contribution to P' again (A.11) but with X_0 replaced by

$$X_n = \frac{(\lambda N(f - f'))^n}{n!} \exp \lambda N(f' - f). \tag{A.16}$$

The factor $n!$ arises from the restriction of the multiple time integral to chronological order. To obtain the total P' we have to sum these n-extra-jump contributions over all n. This is easy, for

$$\sum X_n = 1. \tag{A.17}$$

The result for P' is just (A.1)–(A.4) with f replaced by f'. This is only as expected, but similar manipulations will be useful below.

Suppose now that the system falls into two noninteracting subsystems, with commuting Hamiltonians A and B, respectively:

$$H = A + B. \tag{A.18}$$

Then the operators U factorize:

$$U(t',t) = V(t',t)W(t',t) \tag{A.19}$$

with V and W constructed like U in (A.5), but with A and B replacing H. Since V and W commute, we can collect together the factors referring to each subsystem in (A.2), with the result

$$P = \exp \lambda L(i - f) \exp \lambda M(i - f) \langle i| F^+ F G^+ G |i \rangle, \tag{A.20}$$

where F and G are constructed like E in (A.4) but with operators of the first and second subsystems, respectively. The integers L and M are the 'particle numbers' of the subsystems:

$$L + M = N. \tag{A.21}$$

At this stage the initial and final times i and f are common to the two subsystems. But by the manipulations described above we can pass from i and f to later initial times, and earlier final times. Moreover, because the jump and evolution operators commute with one another, and have been collected together into separate commuting factors F and G, this can be done independently for the two subsystems. So we can take independent initial times i' and i'', and independent final times f' and f'', for the two subsystems, respectively.

The resulting probability distribution, over jumps in the reduced time intervals, is

$$P(f', f''; \mathbf{x}_m, n_m, t_m, \ldots \mathbf{x}_1, n_1, t_1; i', i'') \mathrm{d}^3\mathbf{x}_1 \ldots \mathrm{d}^3\mathbf{x}_m \mathrm{d}t_1 \ldots \mathrm{d}t_m. \quad \text{(A.22)}$$

where

$$P = \exp \lambda L(i' - f') \exp \lambda M(i'' - f'') \langle i', i'' | F^+ F G^+ G | i', i'' \rangle. \quad \text{(A.23)}$$

The jumps and evolutions before i' and i'', in the two subsystems, respectively, have been incorporated into the initial state $|i', i''\rangle$. The jumps and evolutions in the reduced intervals, i' till f' and i'' till f'', make F and G, as in (A.4).

Note finally that if we are interested only in what happens in subsystem 1, we can sum over all possibilities for the second system in a now familiar way. The result is just (A.22), with reference to jumps in system 1 only, and (A.23) without any operator G. It is equivalent to

$$P = \mathrm{trace}_1 F^+ F \rho, \quad \text{(A.24)}$$

where the trace is over the state space of system 1, and

$$\rho = \mathrm{trace}_2 |i', i''\rangle\langle i', i''| \quad \text{(A.25)}$$

with the trace over the state space of system 2.

References

Bell, J. S. (1981) *J. de Physique* **42**, c2, 41–61
Einstein, A. (1949) *Reply to criticisms. Albert Einstein, Philosopher and Scientist* (Schilpp, P. A. ed.). Tudor
Einstein, A., Podolsky, B. and Rosen, N. (1935) *Phys. Rev.* **47**, 777
Ghirardi, G. C., Rimini, A. and Weber, T. (1986) *Phys. Rev.* **D 34**, p. 470
Schrödinger, E. (1926) *Annal. Phys.* **79**, 489–527
Schrödinger, E. (1935a) *Naturwissenschaften* **23**, 807–12, 823–8, 844–9
Schrödinger, E. (1935b) *Proc. Camb. Phil. Soc.* **31**, 555–63
Schrödinger, E. (1936) *Proc. Camb. Phil. Soc.* **32**, 446–52
Schrödinger, E. (1952) *Brit. J. Phil. Sci.* **3**, 109–23, 233–47
Schrödinger, E. (1957) *What is a law of nature? Science Theory and Man*, pp. 133–47. Dover